Best fishes'
LaOnda Rose

D1122273

Petersburg, Alaska.

SALMON, DESSERTS & FRIENDS

SALMON, DESSERTS & FRIENDS
LADONNA GUNDERSEN

PHOTOGRAPHS BY
OLE GUNDERSEN

PUBLISHED BY

LaDonna Rose Publishing
P.O. Box 1200
Ward Cove, Alaska 99928
www.ladonnarose.com
ladonna@ladonnarose.com
www.facebook.com/ladonnarosecooks

Book design by LaDonna and Ole Gundersen
Prepress and technical assistance by Vered R. Mares, Todd Communications

First printing March, 2011
Second printing June, 2011
Third printing December, 2011
Fourth printing September, 2014

ISBN: 978-1-57833-523-7

Printed by Everbest Printing Co. Ltd. in Guangzhou, China,
through **Alaska Print Brokers**, Anchorage, Alaska.

LaDonna Rose Publishing

P.O. Box 1200
Ward Cove, Alaska 99928
www.ladonnarose.com
ladonna@ladonnarose.com
www.facebook.com/ladonnarosecooks

Distributed by
Todd Communications
611 E. 12th Ave.
Anchorage, Alaska 99501
Phone: (907) 274-TODD (8633) • Fax: (907) 929-5550
with other offices in Fairbanks and Juneau, Alaska
sales@toddcom.com • WWW.ALASKABOOKSANDCALENDARS.COM

THEY THAT GO DOWN TO THE SEA IN SHIPS, THAT DO
BUSINESS IN GREAT WATERS; THESE SEE THE WORKS OF
THE LORD AND HIS WONDERS IN THE DEEP.   psalms 107: 23-24

TO MY HUSBAND OLE;
THANK YOU FOR ALWAYS BELIEVING IN ME
AND STANDING WITH ME IN EVERYTHING I DO.

AND TO SYDNEY ROSE;
FOR BEING THE HOPE IN MY HEART.

ACKNOWLEDGMENTS

I owe boundless gratitude to my family and friends who shared their time, energy and talent with me throughout the writing of this book.

I wish to thank my fishermen friends who opened their lives and hearts in the past and present to me.
Your adventuresome spirit is truly something to be admired.

Thank you to Ellie Duree, for her tremendous support, understanding, assistance and much needed motivation.

A special thank you to my parents Butch and Marian Mott, friends and neighbors who lined up at the door with enthusiasm and excitement nearly every day while I was testing my recipes. You fueled me with ideas and inspiration.

My in-laws Eric and Kay Gundersen, my sister-in-law Chris Mott and Ken and Lynora Eichner, thank you for your vital encouragement.

Words alone cannot express the thanks I owe to Ole Gundersen, my husband, and the co-author of my life, who without complaint, supported me throughout this entire project. He took all the wonderful photos in this book and even washed a pot or two. This book would not have been possible without his vision.

CONTENTS

MORNING BITE breakfast and brunch

SAILOR'S DELIGHT lunch

MUG UP little bites

CATCH OF THE DAY main dishes

GALLEY TREATS dessert

Welcome to Alaska's 1st City
KETCHIKAN
The Salmon Capital of the World

INTRODUCTION

The great thing about salmon is they represent summer; cooking and eating salmon is about as good as it gets. Friends, warm weather and the aroma of salmon caramelizing over a barbecue is glorious. Top it off with a wonderful dessert and you have pure bliss.

As a commercial salmon fisherwoman, I am blessed to have wild Pacific salmon readily available and have been eating this majestic fish for over two decades now. Through the years, when giving salmon as a gift, more often than not, people ask,

"What do I do with it?" and "How do I eat it?"

My goal with this book is to celebrate the extraordinary wild Pacific salmon and to help you cook salmon at home with pleasure and success. To me, that means being able to put terrific, delicious salmon dishes on my table without devoting a weekend to the preparation and execution of a meal. I wrote this book because I want you to enjoy the dazzling flavors of salmon and to be able to cook it as easily, happily and quickly as you make other favorite dishes for your table. To accomplish this, I focused on everyday ingredients and recipes that would not take too much time and effort. Salmon dishes you can make anytime, anywhere, quickly and easily. It's fun food, nothing too fancy, but it's all deeply satisfying. As you flip through the recipes, you will find classic favorites like Wild Salmon Quiche with Parmesan Crust, Baked Mini Salmon Croquettes with Creamy Dill Sauce and Smoked Salmon Dip as well as fresh ideas that capture the flavors of cuisines around the world, but have become familiar favorites because of their broad appeal. Some are quick and simple to prepare, others are a more labor of love-but all are delicious additions to a cook's repertoire of recipes.

I hope that what you gain from this book will not only be a great collection of recipes, but also a glimpse into the world of salmon fishing. When you venture to your local supermarket and you see the orange-red fillet of wild salmon, you will have a better understanding of how to prepare it and what it took to get it there.

From my galley to yours!

LaDonna

THE STARS OF THE SHOW

PINK OR HUMPY SALMON
Average size is 3 to 5 pounds. Pinks are the smallest and most abundant of all salmon and mainly used for canning. Its delicate texture makes it suitable for dips, casseroles, patties and sandwiches. The name Pink comes from the pale color of their meat. As males approach the spawning grounds, they develop a large hump on their back, hence the nickname "humpy". Pinks run from mid summer until early fall.

SOCKEYE OR RED SALMON
These sleek Ferrari's of the sea average 5 to 8 pounds. Sockeye have a deep red meat, firm texture and separates into smaller flakes, making it attractive for hot dishes, soups and salads. Sockeye is the most versatile of all salmon and is available fresh, frozen, smoked and canned. Sockeye run in the spring to early summer.

KETA, CHUM OR "DOG" SALMON
Average size is 8 to 15 pounds, light in color, soft in texture and separates into large flakes. Often used for smoking and canning, making it especially suitable for appetizers and cooked dishes. Male chums develop large "dog" like teeth as they approach the spawning grounds, giving them the nickname "dog" salmon. Chums run in the mid summer to late fall.

COHO OR SILVER SALMON

Average size is 6 to 15 pounds. Cohos are a very popular sport fish, due to their size and feistiness. The nickname silver comes from their bright shiny skin and tale. The flesh is slightly lighter red than Sockeye, separates into large flakes and is an excellent choice barbecued and in all dishes. Cohos run late summer to late fall.

CHINOOK OR KING SALMON

Kings are largest of the Pacific salmon averaging 10 to 30 pounds, with some weighing as much as 70 pounds. Prized for its superior quality, texture and delicious flavor, making it wonderful for barbecuing, roasting or mixed in a salad. Kings run in the spring.

SALMON ARE ANADROMOUS, MEANING, THEY ARE BORN IN FRESH WATER, MIGRATE TO THE OCEAN TO GROW INTO ADULTS AND THEN RETURN TO FRESH WATER TO SPAWN.

Naturally Wild Alaska

Alaska is one of the most bountiful fishing regions in the world, supplying 50% of the seafood produced in the United States. Alaska commercial fishermen and women are individual small business owners, who brave the rough Alaska seas and risk their lives daily to support their families and to provide you wild, natural seafood.

There are no fish farms in Alaska. Alaska law prohibits all types of fish farming. Alaska salmon swim wild and free in the unspoiled, icy, clear waters of the North Pacific. Alaska salmon feed on the sea's natural marine life, unlike Atlantic farmed raised salmon, which are fed a diet of industrially manufactured feed. This is why there is a compelling taste and texture difference between wild Alaska salmon and pen-reared farmed Atlantic salmon.

Alaska yields nearly six billion pounds of seafood per year and is the only state in the nation whose constitution explicitly mandates that all fisheries are maintained on the sustainable yield principle.

Salmon seasons are regulated on run strength and based on daily data collected from the fishing grounds. This method accurately measures run strength to ensure future stock. Limits are precisely calculated to keep Alaska waters stocked with a continuous ever-replenishing supply of seafood, just as nature intended. Alaska seafood is additive-free and provides healthful natural nutrients.

By asking for wild Alaska seafood, you help support the many fishing families throughout Alaska. Eating wild Pacific salmon instead of farm raised Atlantic salmon is one of the most effective ways to ensure plentiful wild salmon runs well into the future. Your support is greatly appreciated.

FRIENDS DON'T LET FRIENDS EAT FARMED SALMON

HEALTH BENEFITS OF WILD PACIFIC SALMON

You have probably heard a lot about omega-3 fats in the news, but what exactly are they and why should you eat them? At one time, omega-3s were abundant in our diet. Now that we eat more processed and fast foods, most of us have become deficient in these important nutrients.

Wild salmon contains omega-3 fatty acids, which are understood to be important for our health. According to many research studies, omega 3s are vital for brain function, healthy skin and hair and a well tuned cardiovascular and nervous system.

Today, salmon is one of the ultimate health foods and one of the very best low fat protein sources available. A diet high in wild salmon may be your magic bullet for greater health and beauty benefits. When possible, choose wild salmon fresh, frozen or canned. It is superior to farmed-raised Atlantic salmon.

F/V Martina and F/V Dove
Petersburg, Alaska.

TIPS AND TRICKS ON SALMON

SELECTING FISH
When selecting fish, it is important to use three out of your five senses:

SMELL: When buying fish it is extremely important that you smell it to ensure that it is fresh. The fish should smell sweet and ironically, not fishy. If it smells fishy or has other strong odors the fish is past its prime, so it is best not to purchase it.

SITE: It's also best to examine the fish when checking for freshness. Whole fish should have bright eyes and firm flesh; whereas fillets and steaks should be firm and bright looking with no brown spots or discoloration.

TOUCH: When examining fish, it's important to touch them to feel for firmness. Fresh fish is never soft. Instead, it is very firm to the touch.

STORING FISH
Fresh fish can be stored in a refrigerator tightly wrapped for up to two days at the most. To ensure the fish that I prepare tastes the best, I always buy my fish the same day I plan on preparing it. If this isn't possible and the fish needs to be kept cold for more than a few days, be sure to wrap the fish tightly and place it in the freezer until ready to use. To thaw the fish, simply run it under cold water.

HOW TO TIME FISH
There is only one way to cook fish: quickly! Nothing is worse than overcooked fish; it is tough, dry and unappetizing. Even the tastiest sauce won't save fish from being terrible if it's overcooked. One rule to remember and that is fish continues to cook after it has been removed from the heat. It's better to remove it from the heat before the clock says it's done. An extra minute or two on the heat could mean the difference between moist and dry fish. Measure the thickest part of the fish and plan on a cooking time of 8 to 10 minutes per inch of thickness. For example, if the fish or piece of fish measures three inches thick, you'll need to cook it between 24 and 30 minutes. Watch the color change as the fish is cooking. Raw fish is translucent; cooked fish is opaque. The bottom of the fish will begin to turn white. Then watch it carefully and remove it from the heat before it is completely white. Cook the fish just until it flakes by simply inserting a fork into the flesh. When the fish is cooked, a fork will slide into the flesh smoothly with no resistance. The flesh will come apart easily, but will still look moist and be just opaque.

COOKING FISH

Salmon is one of the most versatile foods you can prepare. Its mild flavor and firm texture can be matched with a wide variety of flavors and cooking methods. The most popular cooking methods follow.

SAUTÉING

Using just a bit of olive oil and making sure to preheat the pan to medium-high heat, are the two tips for a perfectly sautéed piece of fish. Be sure not to crowd the fish; cook it in batches rather than overcrowd the pan. The best way to sauté a fillet, is to let the fish cook skin side up undisturbed for 3 to 4 minutes to develop a nice crust. Carefully turn the fish and cook for another 2 to 3 minutes, then remove the pan from the heat and let the residual heat cook the fish.

BAKING

Baking is one of the easiest ways to cook fish. Heat the oven to the correct temperature (for fish it generally ranges from 350° to 400° degrees). Line a baking sheet or shallow baking dish with parchment paper. Fish can be baked whole, in steaks or fillets. Prepare the fish by rubbing extra virgin olive oil all over it, as well as drizzle oil lightly over the parchment lined baking dish. Season with salt and freshly ground black pepper and your favorite seasonings. Bake for 15 to 20 minutes or until the fish test done.

BROILING

Broiled fish can really be delicious, especially if you season the fish well before cooking. Make sure the oven is nice and hot by preheating the oven to 350° degrees. Using aluminum foil, line a baking sheet. Rinse and pat dry the fish and place it on the sheet skin side down. Make sure the fish is 4 to 6-inches away from the broiler and watch carefully.

GRILLING

Fattier salmon grill beautifully. When grilling fish, it's best to stick with whole fillets or steaks. Whole fish and steaks are thicker and hold together better even though they take longer to cook. *Make sure the grill is nice and hot:* It doesn't matter what type of grill you use, the only thing that matters is that the grill is nice and hot. Make sure that your grill is very clean and oil it lightly before adding the fish. Place the fish skin-side up on the grill. Once the fish is placed on the grill, it should immediately sear to lock in the juice and firm the flesh. Let it cook undisturbed for two to four minutes before you flip it. The fish will develop a nice crust and will release perfectly when it's ready to turn. For more delicate fish fillets, using a grill basket will make grilling any type of fish much easier. Just be sure to remove the fish from the basket as quickly as possible so it doesn't stick. One method for cooking thinner fillets on the grill is simply putting a sheet of heavy duty aluminum foil on the grill and cook the fish on that.

THE EASY GALLEY PANTRY

A well stocked pantry takes the stress out of meal planning because it ensures that you have the tools you need to whip up a fantastic meal. This list covers all the non-perishable items you need to make any recipe in this book.

THE SPICE RACK
Ground allspice
Ground cinnamon
Cream of tartar
Ground cumin
Ground curry powder
Cayenne pepper
Chili powder
Garlic powder
Dried dill weed
Dried oregano
Dried rosemary leaves
Ground thyme
Red pepper flakes
Italian seasoning
Paprika
Dried parsley
Sea salt and black pepper

SWEETENERS
Confectioners sugar
Granulated sugar
Brown sugar
Honey
Maple syrup

NUTS AND DRIED FRUIT
Almonds
Cashew nuts
Hazelnuts
Plain peanuts
Pecans
Pine nuts
Walnuts
Dried cranberries
White raisins

ON THE SHELF
Baking powder
Vanilla extract
Fast acting yeast
Cornstarch
Unsweetened cocoa powder
Semisweet chocolate
Coconut milk
Sweetened coconut
Graham cracker crumbs
Instant coffee granules
Coffee-flavored liqueur
Oreo® cookie pie shell
Instant vanilla and chocolate pudding
Peanut butter sandwich cookies
Canned pineapple
Panko bread crumbs
Canned salmon
Canned clams
Canned black beans
Canned creamed corn
Canned jalapeños
Pickled jalapeños
Artichoke hearts
Tomato sauce
Low sodium chicken broth

OIL AND VINEGARS
Extra virgin olive oil
Non stick spray
Rice vinegar
Sherry vinegar
Rice wine
Sesame oil

CONDIMENTS
Capers
Lemon juice
Lime juice
White wine
Hot sauce
Low sodium soy sauce
Prepared horseradish
Thai garlic chili pepper sauce
Thai fish sauce
Sweet chili sauce
Apple butter
Sweet pickle relish
Sesame seeds
Salsa

PASTA AND GRAINS
All-purpose flour
Cornmeal
Basmati rice
Multi-grain rice
Spring roll wrappers
Wonton wrappers
Dried fusilli noodles
Dried rice noodles
Dried egg noodles
Dried linguine pasta
Dried fettuccini
Dried lasagna noodles

TOP 10 GALLEY TOOLS
I CAN'T LIVE WITHOUT

I've been asked many times, "what are the basic cooking items I need for my galley?"
These are a few of my favorite things. I go back to them over and over for most
of my tasks, they help me to be efficient and make cooking fun while at sea.

1. 6-INCH SANTOKU KNIFE This is the work horse of my galley and is good for just about anything. The knife has a rounded tip which helps to protect my fingers when our boat takes an unexpected roll.

2. LARGE AND SMALL CHOPPING MATS Flexible, easy to clean and the knife glides on them. The larger size allows you to chop a variety of items at once without having to put individual items in their own bowl. What I really love about them, is they are fantastic for small spaces and don't take up much room.

3. A FINE TOOTH MICROPLANE GRATER This is the best tool, in my mind, for everything from zesting citrus fruits, to producing a fine shred of chocolate or cheese.

4. PARCHMENT PAPER I first used these when I owned Viking Ave Bakery in Poulsbo, Washington. When I sold the bakery, I bought some for myself. They are fabulous, not much will stick to them and they save so much clean-up time. I always use parchment paper when cooking fish.

5. A SILICONE SPATULA These spatulas are sturdy enough to stir while you are mixing and are great for scraping every last drop of batter out of the bowls. They're also heat-resistant up to 400°, so there's no danger of finding a reshaped kitchen tool when you are working with hot liquids.

6. NON-STICK LOCKING TONGS I use my tongs for everything. I use them to grab potatoes from the oven, taking corn out from boiling water or any task that requires a good grip.

7. GARLIC PRESS This gadget, hands down, is my favorite time-saver, making garlic prep a snap.

8. WHISK I use my stainless-steel whisk for beating eggs and cream, mixing soups, sauces, just about everything.

9. A SMALL-CAPACITY FOOD PROCESSOR This little machine is perfect for chopping nuts and turning fresh herbs into flavorful seasoning pastes. A real time-saver.

10. CAST IRON SKILLET From stove top to oven, they are the most indestructible all purpose pans I have ever used. Once seasoned they are a breeze to clean-up.

HOW DID A CALIFORNIA GIRL END UP
ON A FISHING BOAT IN ALASKA?

I'll never forget my first fishing trip in Southeast Alaska nearly 24 years ago, braving a chilly 2 day September Seine opening. I climbed aboard the 58 foot F/V Nestor with a crew of 5. I quickly found the oil stove in the galley was the only place to stay warm and wondered how in the heck do you cook on such a thing!

Standing on deck I watched my soon-to-be husband Ole jump into the seine skiff that pulled the net off the boat and drove it to the beach. When it was time to bring the net in, I felt like I was on a treasure hunt, the net and it's contents a suspenseful mystery. Some hauls were full of salmon and some were not. I soon forgot how cold I was and the day turned into an exhilarating fun adventure. I had the opportunity to jump into the seine skiff with Ole for one set. After what seemed like an hour turned out to be a few minutes. The skiff was rocking me to sleep, Ole pointed to the engine cover and said, "crawl up there and take a nap". I did and later he said "right then, he knew I was the one".

I spent 2 days on the open sea in Alaska, watching and helping, I even made pancakes on that oil stove! I gained a full understanding of a fisherman's life. Ole and I married that fall and started our own fishing business gillnetting for salmon, pot fishing for shrimp and long lining for halibut. Each fishing trip I took contributed to the respect I have for the profession, for the people who do it and for their families who support them, because it is in their blood and they love the freedom and independence fishing offers.

Ole and I enjoy salmon fishing, we love each other's company and wouldn't trade it for the world.

BEFORE YOU BEGIN

Canned salmon being a natural product is packed with the skin and small bones intact. The skin and the bones are cooked with the meat in the canning process. This makes them soft and edible as they contain much of the nutritive value of the fish. If you prefer using salmon without any skin and bones, you can easily remove them if you wish or purchase the skinless and boneless variety.

A small amount of salt is added during the canning process to bring out the flavor of the salmon. Those of us who prefer to use less salt in our cooking, will find the salt in canned salmon may be all that is needed. Therefore, I recommend salt to taste in many of my recipes.

You will see on some recipes, skinned side up or skinned side down. This refers to the side of the salmon that previously had skin on it.

I tested all the recipes that appear here and many more, under high sea situations and under normal kitchen situations as well, and I chose the ones I love and have used for years. No recipe was included if there were preparation problems. All of my recipes are fit for land or sea and reflect the same results.

During the salmon season I work in a 4 by 7 foot galley without fancy pots and pans, expensive knifes or kitchen gadgets (other than a hand mixer, blender and a mini food processor). Although I use cast iron cookware in my everyday cooking, nonstick cookware works equally well.

I hope you enjoy this book of mine. I hope it will live in your kitchen, bringing you joy and making you smile. I hope you will enjoy the salmon dishes you make and the time you spend making it. I appreciate and thank you for allowing me to share my world with you.

The F/V Mindalina

MORNING BITE

It is remarkable how this first meal can set you on the right course, like a compass pointing you in the right direction. Breakfast gives your body the fuel it needs to be at it's peak, jump starting your brain to feel alive and aware.

TOASTED BAGEL WITH SMOKED SALMON AND FRESH TOMATOES

WILD ABOUT SALMON

HERB CRÊPES WITH SMOKED SALMON AND LEMON ZEST

FLUFFY SALMON OMELETTE

WHIPPED EGGS WITH PINK SALMON

DELIGHTFUL SALMON BREAKFAST QUICHE

EGGS AND SMOKED SALMON IN A PUFF PASTRY

CHEDDAR AND SMOKED SALMON MINI FRITTATAS

EASY SALMON AND CHEDDAR STRATA

TROLLING  PRIMARY CATCH
KING AND COHO SALMON

A troller is a fishing boat, typically 30 to 58 feet in length, and is a one to two person operation. Two to four long poles extend out to each side to separate the gear. The poles have several lines the fishermen will snap baited hooks and lures to and slowly trails them through the water behind the boat.

Sometimes the poles will have bells at the end to alert the fishermen when a salmon is caught. The fishermen will stun the salmon while it's still in the water, so when he pulls it aboard it's not flopping around to bruise the meat or loose its scales. The salmon, once cleaned, are packed on ice or frozen right away. Troll caught salmon are given great care and quality is very high.

TOASTED BAGEL WITH SMOKED SALMON

As a tempting alternative to the classic lox and bagels with cream cheese, I've put together an unbelievably good combination, that could easily be doubled or tripled.

INGREDIENTS
2 tablespoons cream cheese
2 ounces smoked salmon
4 red onion slices
4 tomato slices
2 toasted whole wheat bagels
(I like everything bagels)

To serve, smear toasted bagel bottom with cream cheese. Place salmon on top, then onion, tomatoes and bagel top.

MAKES 2 SERVINGS

Tree point light house. The first and only light house to be built on mainland Alaska. The light was activated on April 30, 1904.

WILD ABOUT SALMON

This recipe is dedicated to my Alaskan friends, who truly know what this picture represents.

INGREDIENTS

1 (6-to 7-ounce) jar or can salmon

Salmon is a fabulous breakfast treat along with a toasted bagel and fresh blueberries.

SERVES 1

HERB CREPES WITH SMOKED SALMON AND LEMON ZEST

FOR THE CRÊPES
2 eggs
1 ¼ cups milk
1 cup all-purpose flour
2 tablespoons olive oil
¼ teaspoon salt
1 tablespoon fresh chives, finely chopped
1 tablespoon fresh dill, finely chopped
nonstick spray

FOR THE FILLING
1 (8-ounce) package cream cheese, softened
3 green onions, minced
2 tablespoons fresh dill, chopped
1 tablespoon lemon juice
1 tablespoon sour cream
1 teaspoon grated lemon zest
¼ teaspoon black pepper
1 pound smoked salmon (about 2 cups)
1 cup fresh baby spinach

This is a quick and simple way to wow your family!

THE CRÊPES
Blend eggs, milk, flour, olive oil and salt in a blender until smooth. Add chives and dill and pulse 1 or 2 times to just combine. Chill batter, covered 30 minutes.

THE FILLING
Blend cheese, onions, dill, lemon juice, sour cream, lemon zest (grated lemon peel) and pepper in a bowl with a mixer until smooth. Set aside.

Stir batter to redistribute herbs. Heat a 10-inch nonstick skillet over medium heat. Spray bottom and sides with nonstick spray. Holding skillet off heat, pour in ¼-cup batter, immediately tilting and rotating skillet to coat bottom. (If batter sets before skillet is coated, reduce heat slightly for next crêpe.) Return skillet to heat and cook until crêpe is just set and pale and golden around edges, 10-15 seconds. Loosen crêpe with a heat resistant plastic spatula, then flip crêpe over gently with your fingertips. Cook a few seconds longer. Transfer crêpe to a wire rack to cool. Make other crêpes in the same manner, using nonstick spray for each one.

To serve, spread 2 tablespoons of the cheese filling on to top half of each crêpe. Top with 2 tablespoons salmon, then a few spinach leaves. Fold bottom half of crêpe up; fold again to make a triangle shape. Repeat with remaining crêpes and serve.

MAKES 8 CRÊPES

A Pictograph is a design painted on a rock with pigments made by mixing grease or salmon eggs with charcoal, clay or other minerals. Pictographs are most often found in sheltered rock overhangs, high above the water's edge. This pictograph was taken on a voyage into Behm Canal, surrounded by the pristine beauty of the Misty Fjords.

FLUFFY SALMON OMELETTE

INGREDIENTS
1 (14-to 15-ounce) can salmon, drained
6 large eggs, separated
3 tablespoons butter
¾ cup mushrooms, button or cremini,
   sliced
2 green onions, chopped
½ cup Swiss cheese, grated
salt and freshly ground black pepper

Eggs are easiest to separate when they are cold, but will beat to greater volume when they have been allowed to warm at room temperature. Even a tiny speck of yolk will retard the white's expanding, so do separate eggs carefully. A non-plastic, grease -free bowl will help whites to achieve the greatest volume. For brunch, add some sliced fruit and a coffee cake and you're set.

Preheat the oven to 350°F.

Melt 2 tablespoons butter in a small skillet over medium heat. Sauté the mushrooms and onions, set aside to cool.

In a large bowl using a mixer, beat egg whites until stiff. In a separate bowl beat egg yolks, season to taste with salt and pepper.

Fold yolks into the beaten egg whites. Gently add the mushrooms, onions, cheese and salmon.

Melt 1 tablespoon of butter in a 10-inch ovenproof skillet, swirl around edges of pan. Pour omelet mixture into the skillet and spread evenly. Cook over low heat 3-5 minutes to brown the bottom.

Bake 15-20 minutes until golden on top and a knife inserted near the center comes out clean. Serve right away.

MAKES 4 SERVINGS

Wild flowers in the sunset.

WHIPPED EGGS WITH PINK SALMON

INGREDIENTS
1 (6-to 7-ounce) can salmon, drained
6 eggs
⅓ cup milk
salt and freshly ground black pepper
1 tablespoon butter
1 tablespoon fresh parsley, chopped
toasted bread for serving

When you are pressed for time, try this super simple and satisfying breakfast.

Break eggs into a medium sized bowl; add milk, season to taste with salt and pepper and whisk until combined.

Melt butter in a medium skillet over medium heat. When hot add the eggs. When they begin to thicken add the salmon, stirring gently until eggs are set.

Garnish with parsley and serve right away with toasted bread and fresh fruit.

MAKES 2-4 SERVINGS

Wildlife often compete for the same piece of real-estate.

DELIGHTFUL SALMON BREAKFAST QUICHE

INGREDIENTS
1 (14-to 15-ounce) can salmon, drained and liquid reserved
12 eggs
⅛ teaspoon hot sauce
½ cup all-purpose flour
¼ teaspoon salt
1 teasoon baking powder
2 cups Monterey Jack cheese, grated
2 cups cottage cheese
1 cup fresh spinach, chopped

This salmon breakfast quiche could be called a lot of things; breakfast casserole, frittata, crustless quiche, but no matter what you call it, you'll agree this recipe is a deliciously different brunch or breakfast item. Ole lived on these while crab and black cod fishing off the Oregon coast. Left overs freeze beautifully.

Preheat the oven to 350°F.

Whisk the eggs in a large bowl until well blended, add the reserved salmon liquid and hot sauce.

Sift the flour, salt and baking powder. Blend into the egg mixture. Add the salmon, cheeses and spinach, mix well. Pour into a well greased 9 by 13-inch baking dish.

Bake for 35-45 minutes or until a knife inserted near the center comes out clean. Serve right away.

MAKES 6 SERVINGS

EGGS AND SMOKED SALMON IN A PUFF PASTRY

INGREDIENTS

1 frozen puff pastry sheet, thawed
8 eggs
½ cup milk
¼ teaspoon salt
¼ teaspoon black pepper
1 tablespoon butter
4 ounces cream cheese
2 tablespoons green onion, minced
1 teaspoon fresh dill, chopped
3 ounces thinly sliced smoked salmon
⅓ cup Mozzarella cheese, grated
1 egg, slightly beaten
1 tablespoon water

Preheat the oven to 350°F.

Line a baking sheet with parchment, set aside. Whisk the eggs in a medium bowl until well blended, add the milk, salt and pepper.

In large skillet melt butter over medium heat, pour in the egg mixture. Cook without stirring until mixture begins to set on the bottom and around the edges. Using a spatula, lift and fold the partially cooked eggs so the uncooked portion flows underneath. Continue cooking until eggs are just set. Remove from heat. Dot with cream cheese and sprinkle with green onion and dill. Stir gently until combined.

Unfold pastry on a lightly floured surface. Roll into a 15 by 12-inch rectangle. Place on baking sheet. Arrange the smoked salmon crosswise down the center ⅓ of the pastry, to within 1-inch of the top and bottom. Spoon the eggs over salmon and sprinkle with the Mozzarella cheese.

Combine beaten egg with water. Brush the edges of the pastry with the egg mixture. Fold one short side of the pastry over filling. Fold remaining short side over top. Seal top and ends well and brush the top of the pastry with the egg mixture. Top with 10-12 puff pastry stars, if desired and brush with egg mixture. Bake 25 minutes or until pastry is a lightly-golden brown.

MAKES 6 SERVINGS

This pastry is perfect for an afternoon brunch and is usually met with gasps of awe and excitement as it appears, swiftly followed by virtual silence with only the squeaking of forks on plates can be heard.

CHEDDAR AND SMOKED SALMON MINI FRITTATAS

INGREDIENTS
**1 (6-to 7-ounce) can smoked salmon,
drained**
12 large eggs
½ cup onion, chopped
¼ cup green bell pepper, chopped
1 tablespoon fresh dill, chopped
2 cloves garlic, minced
½ cup Cheddar cheese, grated
salt and freshly ground black pepper
muffin pan
nonstick cooking spray

*This is my favorite contribution when we are invited to a dock party.
My version of frittatas disappears fast. Simply scrumptious and easy
to make. I like to use a soup ladle to divide the mixture into the
muffin pan.*

Preheat the oven to 350°F.

Whisk the eggs in a large bowl; add the onion, bell pepper, dill and garlic. Stir
in the salmon and cheese, season to taste with salt and pepper.

Spray the muffin pan and divide the egg mixture evenly.
Bake for 15-20 minutes or until a knife inserted in the center comes out clean.

MAKES 12 SERVINGS

Downtown Meyers Chuck, Alaska.

EASY SALMON AND CHEDDAR STRATA

INGREDIENTS
8 slices whole grain bread
¼ cup butter, softened
8 large eggs
1 teaspoon brown sugar
1 tablespoon Dijon mustard
2 ½ cups half-and-half
3 cups Cheddar cheese, grated
1 pound smoked salmon
    (about 2 cups), skin and pin
    bones removed
½ teaspoon paprika

This is a wonderful recipe for impressing the in-laws. Just put it together the night before, then pop it into the oven the next morning. You'll make your honey proud when his mom asks for the recipe!

Butter a 9 x 13 inch baking dish. Butter bread on both sides, stack the bread and cut it into small cubes.

In a medium sized bowl: whisk the eggs, brown sugar, Dijon and the half-and-half.

Layer one half of the bread in the prepared pan, one half of the cheese and one half of the salmon. Repeat layers. Pour the egg mixture over all layers. Sprinkle paprika over the top. Chill overnight.

Remove Strata from refrigerator ½ hour before baking. Preheat the oven to 350° F. Bake for 45 minutes to 1 hour or until a knife inserted near the center comes out clean

Variation Tip: *Fresh cooked salmon or canned salmon may be substituted for the smoked salmon in this recipe.*

MAKES 4 TO 6 SERVINGS

A deckhand's work is never done.

The F/V *Shanimah* gillnetting for salmon

SAILOR'S DELIGHT

Lunch is an important meal on a fishing boat, but for busy families, it a can be an afterthought. With these quick and easy lunches, you'll find fresh and tasty ideas your whole family will love.

VINTAGE-OF-THE-SEA-CHOWDER

CRANBERRY-ALMOND SALMON SANDWICH

SALMON DILL CROISSANT

WILD SALMON QUICHE WITH PARMESAN CRUST

MUSHROOM, SALMON AND WILD RICE SOUP

SMOKED SALMON CLUB SANDWICH

FRESH VEGETABLE AND SALMON SALAD

BLACK BEAN AND SALMON TOSTADAS

SASSY SALMON BURGERS WITH SAVORY MINT SAUCE

SESAME SALMON WITH SUGAR SNAP PEAS

SMOKED SALMON EGG SALAD ON EASY GRILL BREAD

HERB FUSSILLI AND SALMON SALAD

LEMON SALMON BURGER WITH CREAMY BASIL SAUCE

AVOCADO, ARUGULA AND SMOKED SALMON SALAD

ROASTED GARLIC, POTATO AND SMOKED SALMON SOUP

DELICIOUS SALMON, CORN AND POTATO CHOWDER

SPICY WILD SALMON SANDWICH

GILLNETTING PRIMARY CATCH
SOCKEYE, CHUM AND COHO SALMON

A gillnetter is a fishing boat that is 32 to 45 feet in length and is a one to two person operation. A southeast Alaska gill net is 1,200 to 1,800 hundred feet long, about a quarter mile. The meshes of most gill nets are 5 inches in diameter and are easily broken by a slight tug of the hand. A floating cork line holds the net up vertically with a weighted "foot rope" (or lead line) along the bottom. The boat is attached to one end of the net and a big orange buoy ball at the other end. This enables passing ships and other fishermen to see the net.

The concept of gillnetting is the fisherman sets his net and tries to intercept the salmon as they come in off the ocean heading to their mother streams. The theory is the salmon swim into the net and are caught by their gills. The net is reeled aboard hydraulically and the fisherman will remove the salmon one by one and immediately put them below deck into the fish hold that contains refrigerated seawater.

A DAY IN THE LIFE

The day starts around 1 am with the roar of the engine and the rattle of the anchorchain. As we ease out of the harbor, the engine purrs me back to sleep. Moments later I feel the rise and fall of the ocean, as we make our way to the fishing grounds. In the darkness to the east is the bright glimpse of sunrise. When we arrive at the fishing spot, Ole makes a decision depending on which way the tide is flowing, where to set our net.

(The net is set in the dark to take advantage of the *morning bite*).

As I slumber with our boat cat Sobe wrapped tightly by my side, I feel a gentle tap or a shout that I am needed on deck. Extenuating circumstances such as a very large set of salmon or an unpredictable tide pattern that is pushing us into the rocky shoreline or both has caused me to loose my much needed beauty sleep!

I jump out of bed and put my rain gear on like a fireman at sea, moving as fast as I can! Heavy socks, boots, rain pants, coat, hat and gloves. Buttoning up my jacket, I look around me and see the distant lights on the horizon from other fishing boats. Ducking under the jelly fish ridden net, I join Ole on deck. He tells me the tide is moving twice as fast as he thought and we need to get picked up before we get pushed into the shallows. I take my place opposite him and working together we pick the fish from our net as fast as we can. When the last fish is pulled from our net, Ole immediately starts planning where to set the net again. I take all my rain gear off and hang my gloves above the oil stove to dry. About the time Ole finishes setting the net, I have breakfast prepared. If the weather is nice enough, we will share breakfast up on the flying bridge, which is the top deck of our boat. Watching our net very carefully for logs, seaweed and kelp patches because, there's nothing like the memory of spending four hours picking kelp and pop weed from a gill net to remind you to keep your eyes open. When breakfast is done, it's time to haul the net in. Depending on the amount of fish, I will either finish clean-up or help Ole on deck.

This process of setting and hauling the net repeats itself throughout the day. The only change of pace is when it's time for us to head to the tender, to off load our *catch of the day* (24 hour period). As we leave the tender to head back to the fishing grounds, I make us a little snack we call *Mug Up*, something to hold us over until dinner. We finish our day of setting and hauling around 9 pm as the sun is beginning to set. Often times I will have prepared us a *galley treat* for our hard day's work. The anchor goes down and the roar of the engine stops. We head to bed for a few short hours before it starts all over again.

As the days of summer begin to shorten, our sleep increases until we are waking up at 5 am and heading to the harbor at 6 pm for an early dinner and a movie before bed. To some this lifestyle may seem daunting-the commercial salmon fishing season is open a short 14-15 weeks each year. We work hard during this time, knowing the season will come to an end. We have fun with it and enjoy the time we spend together working side by side.

On our way to the tender.

VINTAGE-OF-THE-SEA CHOWDER

INGREDIENTS

- 4-5 large potatoes, peeled, diced
- 5 thick slices of bacon, cut into ½ inch pieces
- 5 tablespoons butter
- ¾ cup onion, chopped
- 4 cloves garlic, minced
- ¾ cup red bell pepper, diced
- ½ cup carrots, grated
- ¾ cup celery, diced
- ¾ cup all-purpose flour
- 6 cups water
- 2 (6-ounce) cans clams with juice
- 2 pounds (3-4 cups) assorted seafood, salmon, halibut, crab, shrimp
- ¾ teaspoon salt or to taste
- ½ teaspoon black pepper
- 1½ teaspoons curry powder
- 1 cup half-and-half
- 1 cup milk
- 1 tablespoon fresh parsley, minced

Plenty of vegetables make this a hearty chowder your whole family will enjoy. To make a meal add a fresh green salad and a loaf of buttered sourdough bread.

Boil potatoes until tender, drain and reserve.

In a soup pot, fry the bacon until crisp. Using a slotted spoon, transfer to paper towels to drain. Add butter to the soup pot. Add the onion, garlic, red bell pepper, carrots and celery, sauté over medium heat until soft.

Stir in the flour. Pour in the water and clams with juice. Bring to a slow boil, stirring frequently, until thickened. Add the seafood, potatoes, bacon, seasonings, half-and-half and milk. Simmer until all of the fish is opaque throughout, another 5 minutes.

Ladle into warmed soup bowls and garnish with the parsley.
Serve immediately.

Note: *Yukon gold potatoes are an excellent choice for this recipe because of their buttery taste and firm texture. Classified as all purpose potatoes, they are lower in starch than russets and hold their shape well during cooking.*

Variation Tip: *Canned salmon may be used in place of the fresh Salmon.*

MAKES 4 SERVINGS

Dungeness crab.

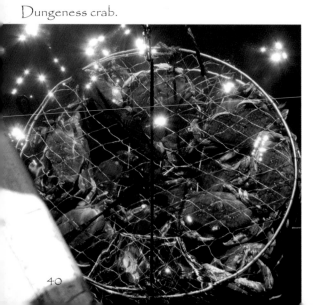

Meyers Chuck, Alaska.

CRANBERRY-ALMOND SALMON SANDWICH

This is fast and easy to make. You will love the sweetness from the cranberries and the crunchiness from the celery, carrots and almonds.

INGREDIENTS
1 (6-to 7-ounce) can salmon, drained
1-2 tablespoons mayonnaise
2 teaspoons dried cranberries
1 tablespoon celery, diced
1 tablespoon carrots, shredded
2 teaspoons slivered almonds
pinch of cinnamon, optional
4 slices whole-wheat bread

Combine the salmon, mayonnaise, cranberries, celery, carrots, slivered almonds and a pinch of cinnamon in a bowl.

Sandwich between two slices of bread and serve at once.

MAKES 2 SERVINGS

Misty Fjords waterfall.

SALMON DILL CROISSANT

INGREDIENTS
- 1 (8-ounce) package cream cheese, softened
- 1 (6-to 7-ounce) can salmon, drained
- ¼ cup mayonnaise
- 1 tablespoon lemon juice
- 1 tablespoon onion, finely minced
- 1 teaspoon prepared horseradish
- ½ teaspoon dried dill
- ⅛ teaspoon salt or to taste
- ¼ teaspoon garlic powder
- 6 croissants, split
- lettuce leaves

These salmon sandwiches taste so wonderfully rich, that every time I serve them everyone assumes they're gourmet. No one needs to know they're fast and easy to prepare by using canned salmon. This is also tasty on crackers.

In a mixing bowl, beat the cream cheese until smooth. Add the mayonnaise, lemon juice, onion, horseradish, dill, salt and garlic powder. Stir in the salmon.

To serve, spread mixture on each croissant bottom, add lettuce leaves and croissant top.

MAKES 6 SERVINGS

Skiffs are not only used for pleasure, they are a way of transportation.

43

WILD SALMON QUICHE WITH PARMESAN CRUST

INGREDIENTS

- 1 (14-to 15-ounce) can salmon, drained and flaked
- 2 tablespoons butter
- 1 medium onion, chopped
- 2 cloves garlic, minced
- 1 cup sour cream
- 4 eggs
- 1 ½ cups smoked Gouda cheese, freshly grated, divided
- 1 tablespoon fresh dill, minced
- ¼ teaspoon salt
- ¼ teaspoon black pepper

FOR THE PARMESAN CRUST

- 1 ½ cups all-purpose flour, plus extra for dusting
- ½ teaspoon salt
- 1 tablespoon Parmesan cheese, finely grated
- ½ cup chilled unsalted butter, diced
- ¼ cup ice water or as needed (one tablespoon at a time)

This terrific salmon quiche recipe goes together in a matter of minutes. If you bake some potatoes at the same time and serve it all with a simple tossed salad, you'll have a complete meal and not much cleanup afterwards.

THE PARMESAN CRUST

Sift together the flour and salt in a large mixing bowl. Add the Parmesan cheese. Cut the butter into the flour, until the mixture resembles course meal. Drizzle a few tablespoons of ice water into the flour mixture.

Using a wooden spoon, stir gently adding water, a tablespoon at a time, just until it holds together when you press a handful of it into a ball. The dough should be moist, not wet. Avoid over mixing as this will toughen the dough.

Gather the dough into a ball, wrap well and chill for at least one hour. Press rolled out dough into a 9½-inch deep pie dish.

THE QUICHE

Preheat the oven to 375°F.

In a small saucepan, sauté the onion and garlic in butter until soft. Remove from heat.

In a large mixing bowl and using a wire whisk, stir the sour cream and eggs until combined. Gently stir in the salmon, onion, garlic, dill, salt, pepper and 1-cup smoked Gouda cheese.

Pour the salmon mixture into the prepared crust and top with remaining cheese. Bake for about 50-60 minutes, until golden and quiche is set. Let stand 10 minutes before serving.

Variation Tip: *You may substitute Gruyere cheese for the Gouda with equal results.*

MAKES 4-6 SERVINGS

Beach combed fishing float art work.

MUSHROOM, SALMON AND WILD RICE SOUP

INGREDIENTS

- **4 slices bacon, cut into ½ inch pieces**
- **1 medium onion, sliced**
- **1 celery stalk, thinly sliced**
- **1 cup assorted mushrooms such as, button, cremini and chanterelle**
- **2 tablespoons all-purpose flour**
- **¼ teaspoon Dijon mustard**
- **¼ teaspoon dried rosemary leaves**
- **1 cup cooked wild rice**
- **4 cups reduced-sodium chicken broth**
- **1 cup half-and-half**
- **1 (14-to 15-ounce) can salmon**

Salmon shines with wild rice, mushrooms and rosemary in this satisfying soup that's pure Pacific Northwest comfort food. A real treat on a cold and rainy day.

In a soup pot over medium heat, sauté the bacon until crisp. Using a slotted spoon, transfer to paper towels to drain. Add the onion, celery and mushrooms and sauté until soft.

Stir in the flour, mustard and rosemary. Stir in wild rice and broth and bring slowly to a boil. Reduce heat to low, cover and simmer 10 minutes.

Add the bacon, half-and-half, salmon and juice. Simmer uncovered, stirring occasionally until hot.

Ladle into warmed soup bowls and serve right away.

MAKES 4 SERVINGS

SMOKED SALMON CLUB SANDWICH

INGREDIENTS

- 8 thick slices of bacon
- ½ cup sour cream
- ¼ cup blue cheese crumbles
- 1 clove garlic, minced
- 1 tablespoon Dijon mustard
- 1 tablespoon fresh parsley, finely chopped
- 1 green onion, finely chopped
- 1 pound smoked salmon (about 2 cups)
- salt and freshly ground black pepper
- 12 slices sandwich bread, toasted
- 8 lettuce leaves, bib or green leaf
- 4 thick tomato slices

This is a scrumptious sandwich your taste buds will fall in love with. Surprise your family with this yummy sandwich for your next picnic!

Fry bacon until crisp, drain on paper towels. While bacon is cooking, in a small bowl combine the sour cream, blue cheese crumbles, garlic, mustard, parsley and green onion.

Layer each of the 4 toast slices with ½ cup smoked salmon, another slice of toast, a mound of the blue cheese spread, 2 crisscrossed slices of bacon, 2 lettuce leaves, 1 tomato slice, a dash of salt and pepper and a third slice of toast. Cut sandwiches into quarters. Enjoy!

MAKES 4 SANDWICHES

Cornelia Marie docked in Ketchikan on its way to the Bering Sea. September 2008.

FRESH VEGETABLE AND SALMON SALAD

INGREDIENTS

1 pint cherry tomatoes, halved
1 ½ cups whole green beans, cut into
    1 inch pieces
1 small zucchini, thinly sliced
3 cups button mushrooms, thinly
    sliced
salad greens
2 (6-to 7-ounce) cans salmon, drained
fresh parsley to garnish

FOR THE DRESSING

¼ cup mayonnaise
¼ cup plain yogurt
2 tablespoons sherry vinegar
salt and freshly ground black pepper

*This fresh from the garden salad is one of our favorites! It is
light and very quick to make.*

THE DRESSING

Combine the mayonnaise, yogurt and vinegar. Season to taste with salt
and pepper. Set aside.

Put the tomatoes, beans, zucchini and mushrooms into a large bowl.
Pour over the dressing.

Arrange salad greens on a serving dish. Add the vegetables and then the
salmon.

Garnish with parsley and freshly ground pepper.

MAKES 4 SERVINGS

BLACK BEAN AND SALMON TOSTADAS

INGREDIENTS

8 (6-inch) corn tortillas
cooking spray
1 (6-to 7-ounce) can salmon, drained
   or fresh cooked salmon
½ avocado, diced
2 pickled jalapeños, thinly sliced
1 tablespoon juice from the pickled
   jalapeños
2 cups romaine lettuce, shredded
¼ cup cilantro, chopped
juice from 1 lime
½ fresh tomato, diced
salt, to taste
1 cup black beans, rinsed
   and drained
3 tablespoons sour cream
2 tablespoons of your favorite salsa
¾ teaspoon ground cumin
2 green onions, chopped
lime wedges

For an easy and fresh weeknight meal, this hits the spot.

Preheat the oven to 400°F.

Place tortillas on a baking sheet and spray with cooking spray; turn and spray opposite side. Bake turning once, for 10-12 minutes, until crisp.

Meanwhile, in a medium-sized bowl using a fork, flake the salmon into pieces. Gently mix with the diced avocado and jalapeños.

In another bowl, toss the romaine with the cilantro, lime juice, tomatoes, pickling juice and salt.

Combine the black beans, sour cream, salsa and cumin in a food processor and blend until smooth. Season to taste with salt. Transfer to a skillet and warm over medium heat.

Dividing the ingredients evenly, top the tostada shells with the black bean mixture, salmon and lettuce.

Finish by sprinkling with chopped green onions. Serve with the lime wedges.

MAKES 4 SERVINGS

Bar Harbor, Ketchikan, Alaska.

49

SASSY SALMON BURGERS WITH SAVORY MINT SAUCE

INGREDIENTS

1 **small cucumber, peeled and sliced**
½ **cup sour cream**
2 **tablespoons fresh mint, coarsely minced**
1 **clove garlic, minced**
1 **teaspoon lemon juice**
salt and freshly ground pepper
1 **pound wild salmon fillet, skin and pin bones removed**
1 **large egg and**
2 **large eggs, hard-boiled and sliced**
⅓ **cup green onions, finely chopped**
3 **tablespoons olive oil**
4 **soft buns, or rolls split and toasted**
12 **fresh spinach leaves**

There is nothing ho-hum about this light refreshing twist on the old salmon burger. This one is sensational.

Sprinkle cucumbers lightly with salt, let stand 10 minutes and then transfer to paper towels to drain. In a small bowl, whisk the sour cream, fresh mint, garlic and lemon juice. Season with salt and pepper. Set aside.

Slice the salmon lengthwise into ¼-inch wide stripes and then cut into ¼-inch wide pieces. In a medium bowl, whisk the raw egg with the green onions and ¼ teaspoon salt. Stir in the salmon.

Heat the olive oil in a large nonstick skillet over medium heat. Add the salmon mixture in 4 mounds; gently press to form four patties. Cook 3 to 4 minutes. Carefully turn patties and cook without turning until done.

Divide the ingredients evenly, top the salmon patty with mint sauce, cucumber, sliced egg, spinach and top half of bun.

MAKES 4 SERVINGS

Captain Ole, at the helm.

SESAME SALMON SALAD WITH SUGAR SNAP PEAS

INGREDIENTS

¼ cup rice vinegar
3 tablespoons olive oil
2 tablespoons soy sauce
1 tablespoon toasted sesame oil
1 ½ teaspoons sugar
1 ½ teaspoons fresh ginger, minced
2 (6-to 7-ounce) cans salmon, drained
1 cup sugar snap peas, sliced
2 green onions, sliced
6 cups romaine lettuce, thinly sliced
4 radishes, thinly sliced
¼ cup fresh cilantro leaves
1 tablespoon sesame seeds
salt and freshly ground black pepper

This dish makes a great summer meal. Cool and refreshing. Add a loaf of your favorite bread and enjoy!

In a small bowl, whisk the vinegar, olive oil, soy sauce, sesame oil, sugar and ginger. Season to taste with salt and pepper.

In a medium bowl, combine 3 tablespoons of the dressing with the salmon, snap peas and green onions.

To serve, divide lettuce among 4 plates. Mound a ½ cup of the salmon mixture in the center of each plate and garnish with the radishes, cilantro and sesame seeds.

Drizzle with the remaining dressing, about 2 tablespoons per salad and season with freshly ground pepper.

MAKES 4 SERVINGS

SMOKED SALMON EGG SALAD ON EASY GRILL BREAD

INGREDIENTS
6 hard-boiled eggs, peeled
2 green onions, finely chopped
½ cup celery, finely chopped
1 tablespoon fresh dill, chopped
¼ cup mayonnaise, plus more for
   grill bread
1 teaspoon sherry vinegar
salt and freshly ground black pepper
1 (6-to 7-ounce) can smoked salmon,
   drained
salt and freshly ground black pepper
sprouts, lettuce or watercress sprigs

FOR THE EASY GRILL BREAD
2 cups all-purpose flour
1 teaspoon sugar
1 teaspoon salt
½ teaspoon fast-acting yeast
½ cup warm water (105˚-115˚)
¼ cup olive oil, plus more for brushing
1 egg, beaten

A great change from the standard egg salad. This is fabulous on the grill bread and tastes like it's from a gourmet deli.

THE EGG SALAD
Place the eggs in a medium bowl and coarsely mash with a fork. Stir in the onions, celery, dill, mayonnaise and vinegar. Season to taste with salt and pepper. Gently fold in salmon. Set aside.

THE GRILL BREAD
In a large bowl, whisk the flour, sugar, salt and yeast. In a small bowl, whisk the water, oil and egg. Stir into flour mixture until a dough forms. Turn dough out onto a lightly floured surface and knead about 2 minutes. Put the dough into a lightly oiled bowl and let rest 10 minutes.

Cut into 8 pieces. Roll each piece out into a 6-inch round. Lightly oil a large skillet and place over medium-high heat until hot. Grill bread rounds one at a time, turning once, until marks appear and bread is cooked through, about 1 minute per side. Keep bread warm wrapped in a cloth.

To serve, spread ½-teaspoon mayonnaise over each grill bread. Spread egg salad over the mayonnaise and top with the sprouts. Fold to close and serve.

MAKES 8

Doing the dishes at 11:00 pm, heading to the harbor.

HERB FUSSILLI AND SALMON SALAD

INGREDIENTS

- **8 ounces dried fusilli**
- **1 red bell pepper, quartered and seeded**
- **1 small red onion, sliced**
- **12 cherry tomatoes, halved**
- **1 (6-to 7-ounce) can salmon, drained**

FOR THE DRESSING

- **⅓ cup olive oil**
- **3 tablespoons sherry vinegar**
- **1 tablespoon lime juice**
- **1 teaspoon mustard**
- **1 teaspoon honey**
- **¼ cup fresh basil, chopped plus extra for garnish**

This makes a fantastic main course for summertime dining or picnicking. It's easy to prepare and doesn't require you to heat up the oven during those long days when the sun is blazing.

Bring a large pot of lightly salted water to a boil. Add the pasta and cook for 8-10 minutes until tender.

Place the bell pepper quarters under a preheated broiler and broil 5-8 minutes, until blackened. Transfer to a bowl and cover, let sweat for about 10 minutes.

Drain pasta and set aside.

Peel and discard bell pepper skins. Slice into strips.

THE DRESSING
In a large bowl, whisk together the olive oil, sherry vinegar, lime juice, mustard, honey and the fresh basil.

Add the pasta, bell pepper strips, onion, tomatoes and salmon. Toss together gently, then divide among serving bowls. Garnish with basil sprigs and serve right away.

MAKES 4 SERVINGS

Code Words for Fish
Fresh ~ the fish has never been frozen.
Fresh Frozen ~ this is fish that is frozen at sea onboard a fishing vessel or at a shore based processing plant within hours of being caught.
Fancy ~ code for previously frozen. In cases when fish has been fresh-frozen and properly thawed, it can be better than fresh. However, it may have been frozen several days out of the water or frozen, thawed and re-frozen. Chances are, if it's called fancy it's not at its best.
Bright or Silver Bright ~ this term refers to salmon before they've entered fresh water to spawn, when they are still "bright" silver and in good condition.

LEMON SALMON BURGER WITH CREAMY BASIL SAUCE

INGREDIENTS

1 **(14-to 15-ounce) can salmon, drained**
2 **eggs**
¼ **cup fresh parsley, chopped**
2 **tablespoons onion, finely chopped**
2 **cloves garlic, minced**
¼ **cup panko bread crumbs**
2 **tablespoons lemon juice**
2 **teaspoons fresh basil, chopped**
½ **teaspoon dried oregano**
⅛ **teaspoon salt**
pinch red pepper flakes
1 **tablespoon olive oil**
4 **soft hamburger buns**
4 **crisp lettuce leaves**
4 **large tomato slices**

FOR CREAMY BASIL SAUCE

2 **tablespoons mayonnaise**
2 **teaspoons lemon juice**
1 **teaspoon fresh basil, minced**

A quick tasty way to serve up canned salmon. Enjoy this with or without the bun and with a salad.

In a medium bowl, combine the salmon, eggs, parsley, onion, garlic, bread crumbs, lemon juice, basil, oregano, salt and red pepper flakes.
Form into four patties.

Heat the oil in a large skillet over medium heat. When the oil is hot, add the patties and cook for about 4 minutes per side or until nicely browned.

THE CREAMY BASIL SAUCE

In a small bowl, combine the mayonnaise, lemon juice and 1-teaspoon basil.

To serve, set the salmon burger on the bottom bun, a dollop of the basil mayonnaise, 1 lettuce leaf, 1 tomato slice and the top bun.
Serve right away.

MAKES 4 SERVINGS

Ketchikan is five hundred miles north of Seattle, is Alaska's "first city" and is the first port of call for many cruise ships.

AVOCADO, ARUGULA AND SMOKED SALMON SALAD

INGREDIENTS
- **1 (5-ounce) package arugula, thoroughly washed**
- **6 radishes, thinly sliced**
- **1 (6-to 7-ounce) can smoked salmon, drained**
- **1 avocado, sliced**
- **¼ cup Parmesan cheese, freshly grated**
- **freshly ground black pepper**

FOR THE DRESSING
- **¼ cup olive oil**
- **3 tablespoons lemon juice**
- **1 teaspoon Dijon mustard**
- **¾ teaspoon sugar**
- **⅛ teaspoon salt**
- **⅛ teaspoon freshly ground black pepper**

This is a really easy to make cool summer treat. I make it when I'm in a hurry and need to put together lunch quickly.

THE DRESSING
In a small bowl, whisk together the olive oil, lemon juice, mustard, sugar, salt and the freshly ground pepper. Set aside.

In a large bowl, gently toss together the arugula, radishes and half the olive oil mixture.

Arrange on a serving platter with salmon and avocado. Serve right away with remaining olive oil mixture, Parmesan cheese and freshly ground pepper.

MAKES 4 SERVINGS

Ole pulling salmon from the net.

ROASTED GARLIC, POTATO AND SMOKED SALMON SOUP

INGREDIENTS

- 1 whole head garlic
- 2 tablespoons olive oil
- ¼ cup onion, diced
- 1 carrot, finely chopped
- 4 cups low sodium chicken broth
- 4 large new potatoes, cut into ½ inch cubes
- ½ teaspoon dried rosemary
- ¼ teaspoon ground thyme
- 1 cup heavy cream
- ½ cup smoked salmon, canned or fresh, cut into bite-size pieces
- salt and freshly ground pepper to taste
- 1 green onion, thinly sliced

This mouth-watering soup will make you crave more!

Preheat the oven to 375°F.

Cut off the top of the head of garlic to expose the cloves. You may need to trim individual cloves along the sides of the head. Brush the cut cloves with 1 tablespoon of olive oil, then nestle the head into a piece of aluminum foil. Bake in oven until the cloves are tender and nicely browned, about 20 minutes.

Remove roasted garlic from the oven, carefully open the foil and allow to cool slightly. When garlic is cool enough to handle, cut the head in half horizontally so that all of the cloves are exposed. Squeeze both halves to release the roasted cloves into a medium bowl.

While garlic is roasting, heat remaining 1-tablespoon olive oil in a large saucepan. Stir in the onion and the carrot and cook, stirring until soft, about 5 minutes. Pour the chicken broth into the saucepan and add the potatoes, rosemary and thyme. Bring the soup to a simmer over medium heat and cook until the potatoes are tender, about 20 minutes.

Remove about ½ of the potatoes from the pot and reserve. Place the roasted garlic cloves into a blender and add the soup, filling the pitcher no more than halfway full. Hold down the lid of the blender with a folded kitchen towel and carefully start the blender, using a few quick pulses to get the contents moving before letting it run. Puree the soup in small batches, until smooth. Pour into a clean pot. Alternately, you can use a stick blender and puree the soup right in the cooking pot.

Stir in the reserved potato cubes, heavy cream and smoked salmon into the pureed soup and bring to a simmer. Serve hot, with a sprinkle of green onion.

MAKES 4 SERVINGS

Fishermen chatting in Wrangell Harbor.

DELICIOUS SALMON, CORN AND POTATO CHOWDER

INGREDIENTS

3 tablespoons butter
¾ cup onion, chopped
½ cup celery, chopped
⅓ cup carrots, grated
3 cloves garlic, minced
3 cups potatoes, peeled and diced
4 cups low sodium chicken broth
¼ teaspoon salt or to taste
¼ teaspoon black pepper
½ teaspoon dried dill
1 (14-to 15-ounce) can salmon or about 1 ½ cups cooked flaked salmon
2 cups half-and-half
1 (15-ounce) can creamed corn
1 tablespoon fresh parsley, minced

A beautiful chowder, delicate and nourishing, with the flavor of Alaskan fishing life. I make this whenever we need a quick one-dish meal that satisfies us all.

In a soup pot melt the butter over medium heat. Add the onion, celery, carrots and garlic, sauté until soft.

Stir in the potatoes, broth, salt, pepper and the dill. Cover and simmer 20 minutes or until the potatoes are nearly tender.

Reduce heat to low and add the salmon with the juice, half-and-half and creamed corn, stirring until hot.

Ladle the chowder into warmed soup bowls and garnish with the parsley. Serve immediately.

MAKES 4 SERVINGS

Sockeye fillets getting ready for the canner.

SPICY WILD SALMON SANDWICH

INGREDIENTS

2 **tablespoons brown sugar**
1 ½ **teaspoons chili powder**
1 **teaspoon ground cumin**
4 **(5-ounce) wild salmon fillets, skin
   and pin bones removed**
1 **tablespoon olive oil**
2 **teaspoons jalapeños, canned or
   fresh, chopped**
½ **cup mayonnaise**
4 **Kaiser rolls or French bread, split
   and toasted**
crisp lettuce leaves
sliced tomatoes

Removing the Pin Bones: *There are
bones other than the back bone and rib
bones in fish called pin bones. You can
find them by rubbing your fingers along
the fish from head to tail. To remove
them, pull them out with tweezers or
needle-nose pliers. Be careful not to grab
the bones too tightly as they are soft and
may break or tear the flesh.*

*To keep sandwiches from becoming dreaded deck food throughout
the long summer salmon season, I came up with this flavorful spicy
salmon sandwich from my husband's love of his salmon hot and
sweet.*

In a small bowl, stir together jalapeños and mayonnaise. Spread on cut sides
of rolls.

On a plate, stir the brown sugar, chili powder and cumin. Press tops of salmon
fillets into spice mixture.

Heat a large skillet over medium-high heat until hot. Add oil, heat until hot. Add
salmon, spice side down, cook 2-3 minutes or until browned. Turn salmon and
cook another 3-4 minutes or until salmon begins to flake. Place salmon on
bottom half of roll, top with lettuce and sliced tomatoes. Cover with top half.

MAKES 4 SERVINGS

*Still smiling after another
20 hour day at the office!*

The F/V Emily Nicole seining for salmon.

MUG UP

Mug ups are like a buffer-zone, a time to relax with a little bite to hold you over until the next big meal. These little bites are fuss free, impressive and special enough to be served as an hors d'oeuvre.

SALMON PARTY ROLL

SMOKED SALMON IN
CUCUMBER BOATS

SUMMERTIME SALMON BRUSCHETTA

BAKED MINI SALMON CROQUETTES
WITH CREAMY DILL SAUCE

CLASSY PARMESAN BASKET

SOFT SPRING ROLLS WITH
SMOKED SALMON AND FRESH BASIL

WILD ALASKA SALMON KABOBS

SMOKED SALMON DIP

MINI SALMON AND HERB QUICHES

SALMON SALAD WONTONS CUPS
WITH GINGER-LIME DRESSING

PURSE SEINER 🐟 PRIMARY CATCH
PINK AND CHUM SALMON

A purse seiner is a commercial fishing boat no more than 58 feet, a crew of 4 to 5, a power skiff and a net that is 250 fathoms (1,500 feet) long.

The theory is a small power skiff is attached to one end of the net and pulls it off the main boat driving it as close to the beach as possible. While it is doing this, the main boat is moving away from the beach, slowly creating a C-shaped "set" around a school of fish. After about 30 minutes, the two boats will come together. The power skiff end is passed to one of the deck hands on the main boat, as well as a tow line that is then hooked to the main boat. The skiff man becomes responsible for keeping the main boat out of the net. The bottom of the net is drawn closed or "pursed", trapping the salmon inside. Once the salmon are hauled aboard, they are transferred below deck into the fish hold containing refrigerated seawater.

SALMON PARTY ROLL

INGREDIENTS

- 1 (6-to 7-ounce) can salmon, drained
- 1 (8-ounce) package cream cheese, softened
- 1 tablespoon lemon juice
- 1 tablespoon prepared horseradish
- 1 tablespoon onion, minced
- ¼ teaspoon salt
- ½ cup pecans, chopped
- 3 tablespoons fresh parsley, chopped

A wonderful party favorite!

Combine salmon the with cream cheese, lemon juice, horseradish, onion and salt. Mix thoroughly. Chill several hours or overnight.

Combine pecans and parsley. Shape salmon mixture into an 8 by 2-inch roll. Roll in nut mixture. Chill well. Serve with crackers.

SERVES 6

Purse seiner making a set for salmon.

SMOKED SALMON IN CUCUMBER BOATS

These are tasty and beautiful little morsels to serve any time of year!

INGREDIENTS
2 medium size cucumbers
12 cherry tomatoes
1 (6-to 7-ounce) can smoked salmon,
    drained and flaked
¼ cup sour cream
1 tablespoon celery, minced
1 tablespoon green onion, minced
1 teaspoon fresh lemon juice
1 teaspoon fresh dill, minced
salt and freshly ground black pepper
paprika

Cut unpeeled cucumbers in ½-inch thick slices. Using a teaspoon; carefully scoop out some of the seeds from one side only. Turn upside down on a paper towel to drain. Cut the cherry tomatoes lengthwise and hollow them out. (I like to use a grapefruit spoon for this.)

In a food processor, purêe the salmon, sour cream, celery, green onion, lemon juice and dill. Season to taste with salt and pepper.

Using a pastry bag and nozzle, pipe mixture into hollows of drained cucumber slices and tomatoes. Sprinkle tops lightly with paprika.

MAKES ABOUT 2 TO 3 DOZEN

SUMMERTIME SALMON BRUSCHETTA

INGREDIENTS

1 **cup tomato, seeded and diced**
1 **cup cucumber, seeded and diced**
½ **cup yellow bell pepper, diced**
1 **teaspoon grated lemon zest**
2 **tablespoons lemon juice**
2 **teaspoons capers**
1 **clove garlic, minced**
½ **teaspoon salt**
¼ **teaspoon black pepper**
1 **pound smoked salmon, skin and**
   pin bones removed
1 **baguette**

These bruschettas can be prepped hours in advance and refrigerated, but are quick enough to make and serve in 15 minutes. The smoked salmon brings rich deep flavor, making a memorable appetizer.

In a medium bowl combine tomato, cucumber, bell pepper, lemon zest (grated lemon peel), lemon juice, capers, garlic, salt and pepper; cover and refrigerate.

Meanwhile, slice baguette into ¼-inch slices and toast in a 325˚F. oven for 7 to 10 minutes.

Drain liquid from tomato mixture. Spoon 1-tablespoon tomato mixture on top of each toasted baguette round. Place a slice of smoked salmon on top.

How to Remove the Skin from a Salmon Fillet:
Place the fillet skin-side down on a cutting board, with the tail end closest to you. Starting at the bottom by the tail, make a small cut through the fillet to the point where you can feel the skin. Turn the knife so the blade is nearly flat. The knife should be between the skin and flesh. Angle the knife slightly down so the edge points toward the skin. Holding the skin firmly with your opposite fingers, carefully slide the knife along the length of the fillet using a back and forth motion (take care not to cut through the skin). Remove and discard the skin.

MAKES 2 DOZEN

BAKED MINI SALMON CROQUETTES WITH CREAMY DILL SAUCE

INGREDIENTS

1 (14-to 15-ounce) can salmon, drained into a measuring cup

milk, add to salmon liquid to measure 1 cup

1 ½ cups panko bread crumbs, divided use

6 tablespoons Parmesan cheese, freshly grated

¼ teaspoon salt

⅛ teaspoon black pepper

¼ cup butter

¼ cup onion, minced

¼ cup red bell pepper, minced

¼ cup celery, minced

⅓ cup all-purpose flour

¾ teaspoon Thai garlic chili pepper sauce

1 tablespoon lemon juice

2 tablespoons green onions, minced

1 tablespoon fresh dill, minced

cooking spray oil

FOR THE CREAMY DILL SAUCE

⅓ cup sour cream

1 teaspoon capers

1 teaspoon lemon juice

1 teaspoon fresh dill, minced

pinch salt

During a busy work-week, salmon croquettes make a great choice. This easy recipe provides nice flavor for seafood lovers. Moist on the inside and crisp on the outside -mmmm!

Flake the salmon into a bowl, set aside. In a small bowl, combine the bread crumbs, Parmesan cheese, salt and pepper, set aside.

Melt butter in a large skillet over medium heat. Sauté the onion, red bell pepper and the celery until soft. Whisk in the flour until smooth. Add the milk mixture, stir until thickened. Remove from heat.

Stir in the salmon, chili pepper sauce, lemon juice, green onions, dill and ½ cup of the bread crumb mixture. Refrigerate 1 hour.

THE CREAMY DILL SAUCE

Whisk the sour cream, capers, lemon juice, dill and a pinch of salt in a small bowl. Refrigerate until serving time.

Preheat the oven to 400°F.

Line a baking sheet with parchment paper. Place the remaining bread crumb mixture in a shallow plate. Roll salmon mixture into 1 ½-inch balls, coat with the crumbs and place 1-inch apart on the prepared baking sheet. Spray the tops lightly with oil. Bake 20-25 minutes until lightly golden brown.

Serve right away with creamy dill sauce.

MAKES 2 DOZEN

CLASSY PARMESAN BASKET

INGREDIENTS
**1 cup Parmesan cheese, freshly
shredded or inexpensive shredded
Parmesan - the kind in the bag works
great.**

**1 drinking glass
2 ounces smoked salmon
mixed salad greens**

*These make a beautiful presentation. They have a lacy, delicate look
to them.*

Heat a 10" skillet over medium-high heat.

Sprinkle shredded Parmesan cheese into a disk shape (using less cheese
around the edges so it looks lacy). When cheese is slightly golden underneath,
carefully remove it using a spatula.

Quickly drape cheese disk over an upside down glass, golden side up and
press to form a bowl.

Allow to cool completely. Cooled baskets can sit several hours.

This recipe produces one basket. Repeat instructions until you have the
number of baskets needed.

Fill basket with mixed greens and top with smoked salmon.

MAKES ONE BASKET

SOFT SPRING ROLLS WITH
SMOKED SALMON AND FRESH BASIL

INGREDIENTS
3 ounces thin dried rice noodles
4 shiitake mushroom caps, sliced
2 tablespoons soy sauce
12 round spring roll wrappers
2 cups thinly sliced smoked salmon
1 cucumber, cut into matchsticks
  (about 1 cup)
2 small carrots, cut into match sticks
  (about 1 cup)
1 medium avocado, diced
1 cup basil leaves, coarsely chopped
1 cup cilantro, chopped
½ cup red onion, minced
½ cup plain peanuts, coarsely chopped

FOR THE DIPPING SAUCE
1 cup rice wine vinegar
1 green onion, minced

Soft spring rolls are a pleasure to see and to eat and simple to make if you set up a small assembly line with all the components ready to roll up. The secret ingredient here is extra pairs of hands, to make the job interesting and fun.

THE SPRING ROLLS
Submerge rice noodles in hot water 8-10 minutes, until soft and clear. Drain and cut noodles into 2-inch pieces, set aside. Toss mushrooms with soy sauce in a small bowl, set aside.

Submerge one spring roll wrapper in a shallow pan or bowl of warm water for about 30 seconds or until softened. Remove it carefully, draining the water. Place it before you flat on the work surface. Arrange the smoked salmon slices, cucumber, carrots, avocado, basil, cilantro, red onion and peanuts in a log shape in center of wrapper. Lift the wrapper edge nearest to you and roll it away from you, up and over the fillings, tucking it in under them compressing everything gently into a cylinder shape, folding ends to seal. Set the roll aside on a platter seam-side down to dry.

THE DIPPING SAUCE
Combine rice wine vinegar and green onion in a small bowl.

Serve spring rolls whole or halved crosswise, with dipping sauce.

MAKES 10-12 ROLLS

Waiting in line at the Tender to off load our catch for the day. (Making a few phone calls.)

WILD ALASKA SALMON KABOBS

INGREDIENTS

1 (15-ounce) can tomato sauce
¼ cup sherry vinegar
3 tablespoons honey
1 tablespoon fresh ginger, minced
½ teaspoon ground cumin
pinch cayenne pepper
3 tablespoons olive oil
4 cloves garlic, minced
1 tablespoon fresh basil, chopped
salt and freshly ground black pepper
3 pounds wild salmon fillets, skin and
   pin bones removed, cut crosswise
   into sixteen ½-inch thick slices

Dazzle your friends at your next barbecue with these flavorful salmon kabobs. A true Alaskan experience!

In a medium saucepan, combine the tomato sauce, vinegar, honey, ginger, cumin, cayenne, 1 tablespoon of the olive oil and half of the garlic. Bring to a boil. Simmer until thick, about 30 minutes. Transfer sauce to a bowl and let cool. Stir in the basil and season with salt and pepper.

In a shallow bowl, combine the remaining 2 tablespoons olive oil with the remaining garlic and season with salt and pepper. Add the salmon and toss well to coat.

Light a grill. Thread the salmon slices onto bamboo skewers. Grill salmon over medium heat, turning once, until cooked through, about 3-4 minutes. Serve salmon with tomato sauce.

Note: *When making kabobs, you need to soak wooden skewers in warm water for at least 20 minutes to keep them from igniting right there. Metal skewers don't need to be soaked, of course, but they do get (and stay) very hot. We prefer to use bamboo skewers because they're inexpensive, they hold up well and are easy to handle right off the grill. Bamboo skewers can go straight into the garbage or right into the fire.*

SERVES 6

We love catching these big fat king salmon!

SMOKED SALMON DIP

INGREDIENTS
- 1 (8-ounce) package cream cheese, softened
- ¾ cup sour cream
- 1 tablespoon lemon juice
- 2 cloves garlic, minced
- 2 tablespoons green onions, minced
- 1 tablespoon fresh dill, minced
- 1 teaspoon prepared horseradish
- ½ teaspoon salt
- ¼ teaspoon black pepper
- 1 (6-to 7-ounce) can smoked salmon drained

This dip also makes a tasty deviled egg filling.

Beat the cream cheese until smooth. Add the sour cream, lemon juice, garlic, green onions, dill, horseradish, salt and pepper. Mix well.

Gently add the smoked salmon. Chill and serve with crackers.

MAKES 2 ½ CUPS

MINI SALMON AND HERB QUICHES

INGREDIENTS

1 sheet puff pastry, thawed
1 (6-to 7-ounce) can salmon, drained
2 green onions, finely chopped
¼ cup half-and-half
⅓ cup Cheddar cheese, shredded
2 eggs, lightly beaten
1 tablespoon parsley, chopped
muffin tin

These delicious mini quiches are a wonderful party dish because they can be made in advance and then simply reheated before serving.

Preheat the oven to 375°F.

Using a 3 inch cutter, cut out 12 rounds of pastry. Press into lightly greased muffin cups. Combine salmon with the onion, half-and-half, cheddar cheese, eggs and parsley. Spoon into pastry cases. Bake for 15 minutes or until golden and eggs are set.

MAKES 12

SALMON SALAD WONTON CUPS WITH GINGER-LIME DRESSING

These are the best appetizers. I take them to parties all the time and there is not a single one left. Everyone thinks I spent a lot of time making them because they look complicated. You'll love this recipe.

Preheat the oven to 325°F.

INGREDIENTS
30 wonton wrappers
1 tablespoon olive oil
2 (6-to 7-ounce) cans salmon, drained
½ cup green onion, chopped
½ cup red bell pepper, diced

FOR THE GINGER-LIME DRESSING
1 tablespoon lime juice
1 tablespoon soy sauce
1 tablespoon rice vinegar
1 teaspoon sugar
1 teaspoon ginger, finely minced
1 teaspoon Thai garlic chili pepper sauce or your favorite hot sauce
1 teaspoon sesame oil

THE WONTONS
Lightly grease 30 mini muffin cups. Place center of wonton wrapper in wells of prepared pan to form a cup, allowing remainder of wrapper to extend above the pan. Bake in the oven 7 to 9 minutes, until crisp and lightly brown.

THE DRESSING
Whisk together all dressing ingredients in a small bowl and stir together well. Set aside.

Place the salmon in a medium bowl, add green onion, red bell pepper and the ginger lime dressing, stirring to combine. Drain off excess liquid.

To serve, spoon mixture into baked wonton cups.

MAKES 2 ½ DOZEN

Ole and me enjoying May Fest in Petersburg, Alaska. (We are leaning up against an iceberg).

The tender Gene S
off-loading the salmon gillnetter F/V Kyrion.

CATCH OF THE DAY

If most days are so busy and you are lucky to get a home cooked meal on the table, here comes salmon to the rescue! Each recipe is absolutely delicious, family-friendly and good for you. You can prepare all of them in less time than it takes to order a pizza. Your family will marvel at how well you managed to put together a fabulous healthy meal in a matter of minutes.

BAKED ALASKA SALMON LOAF WITH GINGERED SALAD

CURRIED SALMON CAKES

APPLE SPICED SALMON

BAKED MEDITERRANEAN SALMON CASSEROLE

TANGY CITRUS GRILLED SALMON WITH SUMMER GREENS

LINGUINE WITH FRESH HERBS, GOAT CHEESE AND SMOKED SALMON

BROWN SUGAR-GLAZED SALMON

SESAME ROASTED SALMON WITH SWEET AND SPICY RHUBARB SAUCE

HAZELNUT-ENCRUSTED WILD SALMON FILLETS

ROASTED SALMON WITH BOK CHOY AND COCONUT RICE

FETTUCCINE WITH SMOKED SALMON, CREMINI MUSHROOMS AND CHIVES

SEARED SALMON WITH WILD BLUEBERRY SALSA

OVEN ROASTED FISH TACOS WITH MANGO-AVOCADO SALSA

TERIYAKI SALMON WITH SHIITAKE MUSHROOMS

GRILLED SALMON STEAKS WITH SOY MAPLE GLAZE

GRILLED SALMON WITH CREAMED PESTO SAUCE

SEARED SALMON WITH RICE PILAF

ROASTED RED PEPPER AND SALMON LASAGNA

TENDER
FISH PACKER

A tender is a large boat that meets the fishermen in a calm cove near the fishing grounds and buys the catch of the day from the fishermen. The tender transports the fish back to the shore-based processing plant. Most fishermen seldom go to town and therefore rely on the tenders for groceries, supplies, ice, fuel and water. This real time saver allows the fishermen to stay on the fishing grounds. The tenders play a vital role in the fishermen's day-to-day activities and they are much appreciated.

AT SEA CRUISINE

Alaska's commercial fishermen and women work in one of the harshest environments. They endure gale force winds, long work days, fatigue and physical stress. Deck hands take care of all the tasks that need to be done on board during a set including: cleaning the deck of seaweed, keeping a sharp eye on the net and surrounding seas, repairing holes in the net, pitching fish into the fish hold and on most boats, cooking the meals.

On a fishing boat being the cook is not the most luxurious job there is. Most people don't want that position, but it's a good way to get and keep a job. When the rest of the crew has a few minutes to kick back and catch a few winks, the cook is often still hard at work either preparing to cook, cooking or cleaning up. Along with all the responsibilities on deck, the cook also has the responsibility of the galley. Trying to turn out a hot meal under adverse sea conditions - with the boat rocking and rolling at times - can be dangerous. The galley stove has protective guards so pots and pans can't slide off. There are also clamps to hold pots on, so they don't move.

As I was writing this, Ole tells me this goofy story. When he was a deck hand on a seine boat out of Ketchikan, the season was about to begin and the boat was in search of a cook. Down the dock comes this young kid looking for a job. The captain asked if he could cook. The kid says "yes I cook great!" The captain says hop aboard. The next morning they head out for the first day of fishing. In the afternoon, the captain asks the new "cook" to whip up the crew something for lunch. The crew eagerly awaited the first meal from the new cook. Ole said up to that day he never heard of a mustard sandwich. The crew were all in disbelief...Really mustard on bread! The cook didn't last long.

We burn huge amounts of calories a day on the water and our appetites are large. The boat is always rocking and every time the boat moves, we move and adjust for the motion of the boat. We are always moving, even when we are sitting.

On a boat, typically all the fresh produce is eaten in the first week, before it all starts going bad. The back-up plan for produce would be to dip into the canned reserves. A little creativity and planning ahead will enable the savvy cook to have plenty of ready-to-eat tasty treats. There's usually not enough time during the peek of the salmon season to eat and finding and fixing something in 15 to 20 minutes can sometimes be a challenge. Most of our meals are salmon and potatoes, salmon and rice, salmon and beans or salmon and vegetables. We eat salmon nearly every day, sometimes twice a day.

With the help of our tenders, we are able to eat cuisine at sea. Once a week we give them a list of grocery items we need, they fax it to the grocery store in town. The store shops for us and delivers the groceries to the tender when they arrive in town from the fishing grounds. The tender leaves town full of fuel, water, ice and our groceries.

Even with this comfort, there have been a lot of times when we have been fishing three-to-four week stretches and are down to our last can of corn. When you are running out of everything, you have got to be creative with your meals. We start mixing and matching different food items. It's always the stuff you don't care for most that's last to go and that's what you end up eating for days and days.

One good thing about being the cook on a boat, even if it's done wrong, you won't hear many complaints unless it's a mustard sandwich!

Baked Alaska Salmon Loaf with Gingered Salad

INGREDIENTS
1 (14-to 15-ounce) can salmon or cooked flaked salmon
1 egg
2 teaspoons lemon juice
¼ cup milk
3 cloves garlic, minced
¼ cup celery, minced
½ cup onion, minced
2 tablespoons fresh parsley, chopped
1 cup soft bread crumbs
½ teaspoon dried dill
½ teaspoon Thai garlic chili pepper sauce
⅛ teaspoon salt
⅛ teaspoon freshly ground black pepper

FOR THE GINGERED SALAD
2½ tablespoons olive oil
2 tablespoons lemon juice
1 small shallot, minced
1 tablespoon honey
1½ teaspoons ginger, finely minced
1 teaspoon grated orange and lemon zest
½ teaspoon sesame oil
salt and black pepper
mixed salad greens
cherry tomatoes, halved

This fish dish is family approved. We all love this salmon loaf along with this colorful gingered salad. It's a good dish to take to pot-lucks as well.

Preheat the oven to 375°F.

In a large bowl, mix all salmon loaf ingredients, including juice from salmon. Place mixture into a well-greased 9 by 5-inch loaf pan. Bake 35-40 minutes, until brown around the edges.

THE GINGERED SALAD
Whisk together all dressing ingredients in a small bowl and stir together well, season to taste with salt and pepper. Drizzle the dressing over salad greens and cherry tomatoes.

Serve right away with salmon loaf.

MAKES 4 SERVINGS

Unloading salmon to the tender.

CURRIED SALMON CAKES

INGREDIENTS
1 (14-to 15-ounce) can salmon, drained
2 teaspoons curry powder, divided use
8 tablespoons tartar sauce, divided use
¼ cup panko bread crumbs, plus more for sprinkling
2 tablespoons fresh ginger, peeled and grated
4 green onions, finely chopped
1 large egg, lightly beaten
salt and freshly ground black pepper
½ red bell pepper, seeded and diced
2 stalks celery, thinly sliced
1 ripe mango, peeled, diced
juice of 1 lime
olive oil for frying

FOR THE TARTAR SAUCE
½ cup mayonnaise
2 teaspoons dill or sweet pickle relish
2 teaspoons onion, minced
1 teaspoon lemon juice
salt and freshly ground black pepper

Combine all ingredients in a bowl. Season to taste.

With simple store bought tartar sauce, these make-ahead patties are the perfect light supper for those nights when you would rather not fuss. If you prefer to make your own tartar sauce, I've included a simple recipe.

In a bowl, combine salmon, 1½ teaspoons curry powder, 1 tablespoon of the tartar sauce, the bread crumbs, ginger, half of the green onions and the egg. Season with salt and pepper. Form into 6 patties and freeze until firm, about 10 to 15 minutes.

Meanwhile, whisk the remaining 7 tablespoons tartar sauce and remaining ½ teaspoon curry powder in a bowl. In a separate bowl, mix 1 tablespoon of the curried tartar sauce with the bell pepper, celery, mango, remaining green onions and lime juice. Season to taste with salt and pepper.

Heat ¼ inch of oil in a nonstick skillet over medium heat. Sprinkle the patties with the bread crumbs on both sides and fry until golden, 2 to 3 minutes per side. Drain on paper towels.

Serve with the mango salad and curried tartar sauce.

Dicing a Mango:
Wash fruit. Lay fruit on the counter, then turn so the top and bottom are now the sides. Using a sharp knife, make a length-wise cut as close to the long, flat seed as possible to remove each side of the fruit. Trim fruit away from the seed. Score each side of the fruit length-wise and width-wise, without cutting through the skin. Using your hand, push the skin up, turning the fruit out. Cut the fruit off at the skin with a knife.

MAKES 4 SERVINGS

A well-used pair of "Ketchikan sneakers".

APPLE SPICED SALMON

INGREDIENTS
4 (6-ounce) wild salmon fillets, skin
and pin bones removed
salt and freshly ground black pepper
¼ cup apple butter
1 tablespoon lemon juice
½ teaspoon allspice
1 shallot, diced (about ¼ cup)
3 tablespoons slivered almonds

I love this recipe because of its unique flavor and ease of preparation.

Preheat the oven to 400°F.

In a small bowl, whisk the apple butter, lemon juice and allspice until smooth.

Place the salmon fillets skinned side down on a parchment lined baking sheet. Sprinkle both sides of each fillet with salt and pepper. Brush the apple butter mixture over the fillets and pour remaining glaze over them. Sprinkle with the shallots and almonds. Bake for 15 to 20 minutes or until fish flakes easily when tested with a fork.

MAKES 4 SERVINGS

Note: *Removing skin from a salmon fillet doesn't have to be a chore, here's an easy tip for removing the skin without any hassle.*

If you are baking the salmon, place a sheet of parchment on a shallow baking pan. Do not use any oil or cooking spray. Season and bake according to taste. This should be about 15 to 20 minutes. Remove salmon from the oven. Slide the spatula between the skin and the salmon fillet. The salmon will easily lift off the parchment and you can move it to the serving plate. (The skin will stick to the parchment).

Cooking at Sea is Part of the Adventure
A galley in a boat is the kitchen aboard a vessel. Space is limited, as are fuel, refrigeration and water, so planning ahead is critical. In the galley of a 32-foot fishing vessel, you get a tiny diesel-burning stove and a tiny oven a little bigger than a piece of paper. A tiny sink will give you barely enough room to wash a cup - large pots and pans will act as wash basins by themselves. The best cooking in a galley is quick, with minimal prep and clean-up.

Baked Mediterranean Salmon Casserole

INGREDIENTS

6 ounces egg noodles, cook al dente
3 tablespoons butter
½ red bell pepper, diced
2 cloves garlic, minced
¾ cup portabella or cremini
  mushrooms, sliced
1 (7-ounce) jar artichoke hearts, sliced
⅛ cup all-purpose flour
1 cup milk
1 cup heavy cream
2 (6-to 7-ounce) cans salmon, drained
3 green onions, thinly sliced
¾ cup Parmesan cheese, freshly grated
¾ cup panko bread crumbs
1 teaspoon Italian seasoning
salt and freshly ground black pepper
nonstick cooking spray

An enticing combination of salmon, egg noodles, garlic, portabella mushrooms and artichoke hearts. The perfect weeknight recipe, that is a snap to prepare. Serve with crusty bread and a green salad.

Preheat the oven to 350°F.

Melt the butter in a large saucepan over medium heat. Add the bell pepper, garlic and the mushrooms, sauté until soft. Stir in the artichoke hearts and the flour and cook for 1 minute. Pour in the milk and the heavy cream. Stir over low heat until the sauce thickens, about 5 minutes. Season to taste with salt and pepper. Remove from heat.

In a large bowl, combine the egg noodles, salmon, green onions, ½ -cup Parmesan cheese and the sauce mixture. Spray a 10-inch round casserole dish with nonstick cooking spray. Add noodle mixture and top with the bread crumbs, Italian seasoning and ¼ cup Parmesan cheese. Bake for 20 minutes or until bubbly and golden.

MAKES 4 SERVINGS

How to Make Panko Bread Crumbs
Panko makes delicious breading for fish or makes a great topping for casseroles.

You'll need 7 slices of white bread, remove the crust. Preheat the oven to 300°F. Carefully push chunks of bread through the shredding disk of a food processor, to make course crumbs. Spread the crumbs on a baking sheet and bake until the crumbs are dry but not toasted, about 8 to 10 minutes. Shake the sheet twice during baking. Immediately remove the bread crumbs from the oven to cool. Once cooled, store in a re-sealable bag in the freezer.

TANGY CITRUS GRILLED SALMON WITH SUMMER GREENS

INGREDIENTS
4 (6-ounce) wild salmon fillets, skin and pin bones removed
2 thick red onion slices
6 cups mixed spring greens
1 cup pineapple, diced
½ cup cashew nuts, coarsely chopped

FOR THE TANGY CITRUS VINAIGRETTE
1 orange and 1 lemon, juiced and zested
1 lime, juiced
1 tablespoon sugar
salt and freshly ground black pepper
red pepper flakes to taste
¼ cup olive oil

The delicious orange-kissed marinade makes this grilled salmon even more of a treat.

THE VINAIGRETTE
Combine juices and zests (grated peel) of the orange, lemon and lime in a small bowl. Add the sugar and seasonings; whisk in the olive oil until slightly thickened.
Prepare the salad greens, pineapple and cashews; chill until ready to serve.

Marinate the salmon in ⅓ cup vinaigrette for 15-20 minutes. Brush onion with some of the vinaigrette and season with salt and pepper.

Preheat a grill or stove-top grill pan to medium-high heat and lightly oil the grates. Remove fish from the marinade (discard marinade), season lightly with salt and pepper. Grill the fillets skinned side up and cook 3 to 5 minutes. Turn fillets over and grill until the fish is just cooked through, about 3 or 4 minutes more. Grill the onions while the salmon cooks.

To serve, toss the greens, pineapple and cashews with some of the remaining vinaigrette. Divide greens among four serving plates, place a fillet of salmon alongside them, then top the fish with rings of grilled onions. Drizzle fish with additional vinaigrette and serve.

Note: If the fish sticks a bit to the grill when flipping, continue cooking a minute or so until they release. To check for doneness, insert a knife or fork into the thickest part of the fillet and separate the flakes. It should be barely opaque in the center - residual heat will continue cooking it to perfection.

Preperation Tip: Always marinate fish in a non-reactive container or use a self sealing, gallon sized plastic bag. The acid in the marinade causes a chemical reaction with metal, giving your fish a metallic taste and discoloring your pan.

MAKES 4 SERVINGS

Full moon.

Linguine with Fresh Herbs, Goat Cheese and Smoked Salmon

INGREDIENTS

- 1 pound linguine pasta
- 8 ounces goat cheese, crumbled
- 6 tablespoons butter
- 3 cloves garlic, minced
- 3 green onions minced
- ½ cup white cooking wine
- ¼ cup flat leaf parsley, chopped
- ¼ cup fresh dill, chopped
- ¼ cup fresh basil chopped
- 2 cups tomatoes, chopped
- 2 (6-to 7-ounce) cans smoked salmon, drained (or fresh smoked)
- salt and freshly ground black pepper

This herb, goat cheese and smoked salmon linguine is one of the quickest pasta recipes you'll ever make-and one you'll want to make again and again.

Bring a large pot of water to boil, lightly salt it, add the pasta and cook until al dente. Drain, reserving ¾ cup of the pasta cooking water.

In a large serving bowl, add the crumbled goat cheese.

In a medium saucepan, melt the butter over medium heat. Add the garlic and green onions, sauté 1 minute. Stir in the wine and cook, about 2 minutes. Add the parsley, dill and basil. Stir in the reserved pasta cooking water.

Add the pasta to the goat cheese. Pour the herb sauce on top, season with salt and pepper and toss. Add the tomatoes and toss gently.

Flake the smoked salmon on top, gently toss in and serve right away.

MAKES 4 SERVINGS

Ole sets his net about 10 times per day.

Brown Sugar-Glazed Salmon

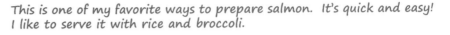

INGREDIENTS

½ **cup butter**
½ **cup brown sugar**
salt and freshly ground black pepper
4 (7-ounce) wild salmon fillets, skin
 and pin bones removed
½ **cup chopped pecans, optional**

This is one of my favorite ways to prepare salmon. It's quick and easy! I like to serve it with rice and broccoli.

Preheat the oven to 400°F.

In a small saucepan; melt the butter and brown sugar. Whisk until combined. Set aside to cool.

Place the fillets skinned side down on a parchment-lined baking sheet. Season with salt and pepper. Brush the butter mixture over the fillets and pour remaining glaze over them. Top with pecans. Bake for 15 to 20 minutes or until fish flakes easily when tested with a fork.

Transfer to warm dinner plates and serve right away.

Freshness Criteria: A whole salmon should look bright, firm and glossy, as though it has just jumped from the water. It should be well cleaned and smell good. Gently press on it with your finger, it should feel firm and the flesh should spring back.

Remember to treat fish as any other meat, following food safety instructions. This means wash your hands, utensils and surfaces with warm soapy water after contact with raw fish.

MAKES 4 SERVINGS

SESAME ROASTED SALMON WITH SWEET & SPICY RHUBARB SAUCE

INGREDIENTS
FOR THE SALMON
- **4 (6-ounce) wild salmon fillets, skin on and pin bones removed**
- **2 tablespoons soy sauce**
- **2 tablespoons sesame seeds**
- **1 green onion, finely sliced**

FOR THE RHUBARB SAUCE
- **2 tablespoons olive oil**
- **3 cloves garlic, minced**
- **1 tablespoon ginger, minced**
- **1 cup rhubarb, thinly sliced tossed with 1 teaspoon sugar**
- **3 green onions, cut into 1- inch pieces**
- **½ red bell pepper, cut into strips**
- **1 teaspoon jalapeños, fresh or canned**
- **2 tablespoons Sweet Chili Sauce**
- **2 tablespoons rice vinegar**
- **2 tablespoons soy sauce**
- **2 tablespoons brown sugar**

Fresh from the garden, succulent rhubarb gives this sauce a lovely tang.

Preheat the oven to 400°F.

Place the fillets on a parchment-lined baking sheet. Brush with the soy sauce and sprinkle with the sesame seeds. Roast for 15-20 minutes or until fish flakes easily when tested with a fork.

THE RHUBARB SAUCE
Heat the oil in a large skillet over medium heat. Add the garlic and ginger, cook for 30 seconds. Add the rhubarb, green onions, bell pepper and jalapeños. Sauté 1 minute, reduce heat to low and add the Sweet Chili Sauce, rice vinegar, soy sauce and brown sugar. Cook to heat through.

To serve, transfer salmon to warm dinner plates, spoon rhubarb sauce over, garnish with green onion and serve right away.

MAKES 4 SERVINGS

HAZELNUT-ENCRUSTED WILD SALMON FILLETS

INGREDIENTS

½ cup hazelnuts
½ cup fresh parsley, chopped
1 tablespoon grated lemon zest
⅛ teaspoon salt and freshly ground
   black pepper
4 (6-ounce) wild salmon fillets, skin
   and pin bones removed
2 tablespoons olive oil
4 cups mixed greens
lemon wedges

Easy enough for family, fancy enough for company. Moist flavorful salmon showcasing the flavors of the Pacific Northwest.

Grind the hazelnuts in a food processor; do not over grind into a paste.

On a plate mix the hazelnuts, parsley, lemon zest (grated lemon peel), salt and pepper.

Dry the salmon with a paper towel, dredge the fillets on both sides in the hazelnut mixture.

Heat the oil in a large skillet over medium-high heat, add the salmon and cook for about 5 minutes on each side until the fish flakes easily when tested with a fork.

Transfer to warm dinner plates and serve with mixed greens and lemon wedges.

MAKES 4 SERVINGS

SALTY SUPERSTITIONS

Fishermen have attributed superstitions to almost all aspects of their work and life on the sea. The job, more so in the past than in the present and in the days of pirates, was a very dangerous one and the slightest mistake could spell disaster for all on board. To deter any type of dangerous action and behavior, superstitions could serve as a warning and therefore minimize future calamities. Part of the romance of the sea is found in its traditions and none have lingered longer than the superstitions of sailors and fishermen.

One of the most repeated verses that originates from a sailor superstition is:

Red sky at night, sailors delight
Red sky in the morning, sailors take warning

For sailors it was lucky:

- to break a bottle of wine or champagne against the hull of a new ship before sailing.
- to step aboard using the right foot first to start the journey off on the right foot.
- for a child to be born on the ship brings good luck to all aboard.
- to throw coins into the sea as a boat leaves port for a safe voyage.
- to have charms such as fox tails, seal skins, shark teeth and sea shells.
- to have a horseshoe on a ship's mast will turn away a storm.
- to toss an old pair of shoes overboard, just as you depart on a journey and good luck follows you on the voyage.
- for dolphins to swim with the ship. Dolphins are considered a sacred friend of fishermen and bring good fortune.

For sailors it was unlucky:

- to begin a voyage on Friday. Most fishermen view Friday as unlucky and would simply refuse to sail. Legend has it that various ships lost at sea departed on a Friday. The most well-known-reason for the dislike, is because it is believed that Christ was crucified on a Friday.
- to have a black "sea bag" aboard brings misfortune. Black bags are considered bad because black is the color of death and a metaphor for the depths of the dark cold sea.
- to throw a stone over a departing vessel dooms the ship and ensures she will never return.
- for the hatch to fall into the fish hold and never turn the hatch up side down. This will cause the hold to fill with sea water.
- to have the ladies aboard. Their presence on a ship--unless they're made of wood and attached to the bow of a ship, was typically thought to be bad luck for everyone aboard.
- to whistle while underway. Whistling is not allowed on many boats-it may have the power to whistle up a storm--a sailor's worst nightmare.
- for the moon to have a ring around it. This is often thought to predict an approaching storm, while a rising moon during a storm means the skies will soon clear.
- to look back once your ship has set sail. Looking back to port implies that you are not truly ready to brave the seas and complete your voyage.
- to bring bananas aboard, a popular theory was that poisonous spiders and other bugs hitched a ride in bananas. And then, of course, there's the theory that banana peels cause crew members to slip and fall on deck.

ROASTED SALMON WITH BOK CHOY AND COCONUT RICE

INGREDIENTS

FOR THE SALMON

- **4 (6-ounce) wild salmon fillets, skin and pin bones removed**
- **¼ teaspoon salt**
- **¼ teaspoon freshly ground black pepper**

FOR THE RICE

- **2 cups uncooked basmati rice**
- **1 ½ cups coconut milk**
- **1 ½ cups water**
- **¼ teaspoon salt**
- **1 cup green onions, chopped**

FOR THE BOK CHOY

- **1 teaspoon olive oil**
- **8 cups bok choy, trimmed and cut into 1 ½ -inch pieces**
- **1 tablespoon fresh ginger peeled, minced**
- **¼ cup rice wine**
- **pinch salt**

FOR THE SAUCE

- **⅓ cup fresh lime juice**
- **¼ cup rice vinegar**
- **2 tablespoons cilantro, chopped**
- **3 tablespoons brown sugar**
- **2 tablespoons Thai fish sauce**
- **½ teaspoon curry powder**

Salmon is a versatile, ultra heart-healthy superfood and its preparation possibilities are endless. Here, a skinless roasted fillet creates a simple but swanky dinner party entrée when paired with coconut rice and bok choy. Just serve with hot jasmine tea and offer sorbet for dessert.

Preheat the oven to 250°F.

THE SALMON

Place salmon fillets skin side down on a parchment-lined baking sheet. Sprinkle both sides of each fillet with salt and pepper. Bake until opaque but still pink in the middle, 25 to 30 minutes. Remove from oven.

THE RICE

Rinse rice with cold water; drain. Combine rice, coconut milk, 1 ½ cups water and ¼ teaspoon salt in a medium saucepan. Bring to a boil over medium-high heat; stir once. Cover, reduce heat and simmer 15 to 20 minutes until liquid is absorbed. Let stand 10 minutes; stir in the green onions.

THE BOK CHOY

Heat oil in a large nonstick skillet over medium-high heat. Add bok choy and ginger; sauté 1 minute. Add rice wine and a pinch of salt; cover and cook 2 minutes or until bok choy wilts. Cover and keep warm.

THE SAUCE

Whisk the lime juice and remaining ingredients in a small bowl, stirring well.

To serve, transfer salmon to warm dinner plates and serve with rice, bok choy and sauce.

MAKES 4 SERVINGS

FETTUCCINI WITH SMOKED SALMON, CREMINI MUSHROOMS AND CHIVES

INGREDIENTS

1 pound fettuccine, cook al dente, drain and set aside
½ cup butter plus 1 tablespoon
4 cloves garlic, minced
1 cup cremini mushrooms, sliced
4 cups heavy whipping cream
1 (6-to 7-ounce) can smoked salmon, drained or fresh smoked salmon
¼ cup fresh chives, minced
salt and freshly ground black pepper
1 tablespoon fresh flat-leaf parsley, minced
¼ cup fresh Parmesan cheese, grated

This is one mouth-watering pasta your family will love!
Serve with Garlic bread.

In a small saucepan, melt 1-tablespoon butter over medium heat. Add the garlic and mushrooms and sauté until soft. Set aside.

Combine the cream and ½-cup butter in a medium saucepan. Cook and stir over medium heat until thick and glossy.

Add salmon, chives, garlic, and mushrooms. Season to taste with salt and pepper. Stir gently for about 1 minute.

Transfer fettuccine to a warm serving platter. Pour sauce over and toss just to blend.

Garnish with parsley and Parmesan cheese and serve.

MAKES 4 SERVINGS

Turn a baguette into a real treat with good extra-virgin olive oil and a fresh garlic rub down!

Cut baguette diagonally to create long slices. Brush each piece liberally with extra virgin olive oil, coating both sides. Toast in a pan as you would cook French toast, carefully browning both sides. Rub one side of toasted bread with a fresh whole clove of garlic. Serve immediately.

Another fresh salmon ready for market.

SEARED SALMON WITH WILD BLUEBERRY SALSA

INGREDIENTS

**4 (6-ounce) wild salmon fillets, skin
and pin bones removed**
olive oil for grilling
salt and freshly ground black pepper

FOR THE BLUEBERRY SALSA

1 cup fresh wild blueberries
**½ cup crushed canned pineapple,
drained**
½ cup red bell pepper, minced
¼ cup pine nuts
¼ cup red onion, minced
¼ cup fresh cilantro, chopped
¼ cup white raisins
1 clove garlic, minced
1 tablespoon lime juice
¼ teaspoon lime zest
1 teaspoon jalapeños, minced
salt to taste

*This special recipe for seared salmon celebrates all the sensual
flavors and freshness of summer. The blueberry salsa provides
beautiful sweet and sour tones to this dish. Be sure to use fresh
blueberries for the best flavor and texture. You can, however, use
frozen.*

THE SALSA
Combine salsa ingredients in a medium bowl. Salt to taste. Cover and
refrigerate one hour.

THE SALMON
Preheat a stove-top grill pan to medium-high heat and lightly oil the grates.
Season the fillets with salt and pepper. Grill the fillets skinned side up and cook
3 to 5 minutes. Turn fillets over and grill until fish is just cooked through, about
3 or 4 minutes more.

To serve, place the salmon on a warmed plate and spoon blueberry salsa on
top.

SERVES 4

*In Alaska, berry picking is akin to beachcombing. It is very addictive. If you've
ever had a blueberry pie made with fresh succulent blueberries, then you know
what I mean. Blueberries and many other berries are all over Alaska. Berry
picking brings out Alaskans in droves to their favorite spots. In Alaska there are
plenty of blueberries to go around and you can pick all you want. Remember,
bears also love blueberries and they have the right-of way. Sing, make noise or
wear bells so they hear you coming!*

*Smoked salmon comes in two distinct varieties: Hot smoked and lox. The
main difference is in the brining process and the temperature during the
smoking process. Hot smoked salmon is fully cooked to 145 degrees, whereas
cold smoked salmon is smoked at 80 degrees and is in fact cured, and still raw,
so needs to be vacuum packed or frozen very quickly after preparation.*

OVEN ROASTED FISH TACOS WITH MANGO-AVOCADO SALSA

INGREDIENTS
FOR THE MANGO-AVOCADO SALSA
1 ripe mango, peeled and diced
½ cup cherry tomatoes, halved
½ cup cucumber, diced
¼ cup red onion, minced
2 teaspoons jalapeño seeded, minced
1 teaspoon sugar
¼ teaspoon salt
¼ teaspoon ground cumin
juice of one lime
1 ripe avocado, pitted, peeled, thinly sliced
3 tablespoons fresh cilantro, chopped

FOR THE CREAMY JALAPEÑO SAUCE
¼ cup sour cream
¼ cup mayonnaise
2 teaspoons sugar
2 teaspoons jalapeño seeded, minced
1 teaspoon lime juice

4 (4-ounce) wild salmon fillets, skin and pin bones removed. Sliced into strips
¼ cup lime juice
¼ cup all purpose flour
⅓ cup yellow cornmeal
1 teaspoon chili powder
½ teaspoon salt
⅛ teaspoon cayenne
8 (7-inch) corn tortillas
3 cups cabbage, shredded

These delicious and easy fish tacos are a big part of our summer rotation! You'll love how quickly they go together.

THE MANGO-AVOCADO SALSA
Combine the mango, tomatoes, cucumber, onion, jalapeño, sugar, salt, cumin and lime juice in a large bowl. Let stand 10-15 minutes to blend flavors. Before serving, gently stir in avocado and cilantro.

THE CREAMY JALAPEÑO SAUCE
Stir all ingredients together in a small bowl and adjust seasonings to taste. Let stand a few minutes to allow flavors to blend. Keep chilled until ready to serve.

Preheat the oven to 425°F. With racks on the top and bottom levels.

Marinate the salmon fillets in lime juice for 5 minutes. Place 2 baking sheets in oven, one on each rack and preheat 5 minutes.

Combine flour, cornmeal and seasonings in a shallow dish. Coat fillets on both sides. Carefully remove the top pan from the oven and spray with nonstick spray. Place fillets on pan and bake for 5 minutes. Spray both sides of tortillas with nonstick spray. Oven-toast tortillas on the second baking sheet after turning the fillets over. Toast tortillas 2 to 3 minutes. Remove fillets and tortillas from the oven.

To make tacos, fill tortillas with shredded cabbage, fish and mango salsa. Drizzle with jalapeño mayonnaise. Serve right away.

MAKES 4 SERVINGS

Teriyaki Salmon with Shiitake Mushrooms and Fresh Chives

INGREDIENTS
- ¼ cup soy sauce
- 2 ½ tablespoons brown sugar
- 2 teaspoons sesame oil
- 4 teaspoons olive oil
- 2 cups shiitake mushrooms, thickly sliced
- 4 (6-ounce) wild salmon fillets, skin and pin bones removed
- 1 tablespoon fresh chives, chopped

This recipe never fails to be a big hit and it is so easy!

Preheat the oven to 400°F.

In a small bowl, whisk the soy sauce, brown sugar and sesame oil. Set aside.

In a large ovenproof skillet, heat 2 teaspoons of the olive oil. Add the mushrooms and cook over medium heat until browned. Add all but 1 tablespoon of the soy sauce mixture and cook, stirring until the skillet is dry and the mushrooms are glazed, about 3 minutes. Transfer the mushrooms to a plate.

Add the remaining 2 teaspoons olive oil to the skillet. Add the salmon fillets skinned side up and cook over medium heat, turning once, until lightly browned. Remove skillet from the heat; pour the reserved 1 tablespoon of the soy mixture over the fillets to coat. Bake the salmon until the top is golden, lightly glazed and just cooked through, 2-3 minutes.

To serve, transfer salmon to warm dinner plates and top with the mushrooms and chives.

MAKES 4 SERVINGS

Bouy ball that marks the end of our net. (This enables passing ships and other fishermen to see our net.)

GRILLED SALMON STEAKS WITH
SOY MAPLE GLAZE

INGREDIENTS

- ¼ cup soy sauce
- 3 tablespoons maple syrup
- 3 tablespoons sesame oil
- 4 (6-ounce) wild salmon steaks
- 2 inch piece of fresh ginger, peeled and thinly sliced
- 2 cloves garlic, minced
- 2 green onions, thinly sliced

Every summer I come up with one new preparation for salmon that is simple and incredibly versatile. This go-to recipe will amaze you and your guests will praise you!

In a large shallow dish, whisk the soy sauce with the maple syrup and sesame oil. Add the salmon steaks and turn to coat. Press ginger and garlic onto both sides of steaks. Cover and refrigerate for 1 hour, turning the salmon a few times.

Preheat a grill or stove-top grill pan to medium-high heat and lightly oil the grates. Remove the salmon from the marinade. Pour the marinade into a small saucepan and boil over high heat until syrupy, about 3 minutes. Strain the glaze into a small bowl.

For good marks, place the salmon steaks at a 45° angle to grates. Cook the steaks 3 minutes. Carefully give them a quarter-turn (45°) and grill 2 minutes. Turn steaks over and grill until just cooked through, about 3 or 4 minutes more.

To serve, transfer to warmed plates and spoon glaze on top. Sprinkle with green onions and serve.

MAKES 4 SERVINGS

The F/V Velvet.

GRILLED SALMON WITH CREAMED PESTO SAUCE

INGREDIENTS
4 (6-ounce) wild salmon steaks
1 cup heavy cream
olive oil for grilling
salt and freshly ground black pepper

FOR THE BASIL PESTO
2 cups fresh basil
1 cup fresh Italian parsley
½ cup Parmesan or Romano cheese, grated
½ cup pine nuts, toasted
4 cloves garlic, roughly chopped
¼ teaspoon salt
⅛ teaspoon red pepper flakes
½ cup olive oil
1 tablespoon lemon juice
2 tablespoons water

Swirling reduced fresh cream in herb pesto creates a delicate sauce for salmon.

THE PESTO
Combine all ingredients in a food processor or blender. Puree until the mixture forms a smooth, thick paste.
Combine 2 cups of the basil pesto and cream in a saucepan. Bring to a simmer and adjust consistency as necessary with additional pesto; season to taste.

FOR THE SALMON
Preheat a grill or stove-top grill pan to medium-high heat and lightly oil the grates. Season the salmon steaks with salt and pepper; lightly coat with oil. Cook the salmon on both sides until fish flakes easily when tested with a fork.

To serve, spoon the sauce on a warmed plate and place the salmon on top.

Note: *Salmon are perhaps the most visually striking of the world's fish. Sleek and silvery, a buttery and succulent flesh. Salmon is the ocean's great natural delicacy. It's neither red-fleshed nor white, making it an exceptionally accommodating wine companion. Depending on the preparation, salmon dishes can be paired with a spectrum of wines including Champagne, Chardonnay and Pino Noir.*

Pesto Variation:
Ginger- Lemon Pesto: Make a regular pesto, adding 1 teaspoon lemon zest (grated lemon peel) and 2 teaspoons finely grated ginger. This pesto is perfect on sautéed fish.

SERVES 4

The F/V Voyager.

SEARED SALMON WITH RICE PILAF

INGREDIENTS
- ½ cup multi-grain rice, soaked for 30 minutes and rinsed
- 1 tablespoon olive oil
- ½ medium onion, diced
- 1 clove garlic, minced
- ¼ cup oyster or cremini mushrooms, sliced
- 1 cup chicken or vegetable broth
- 1 tablespoon slivered almonds
- 1 tablespoon fresh parsley, chopped

FOR THE SALMON
- 2 (6-ounce) wild salmon fillets, skin and pin bones removed
- 1 teaspoon olive oil
- salt and freshly ground black pepper

A quick, healthy and flavorful dish. Excellent for a weeknight dinner and could be dressed up for a delicious dinner party.

THE RICE PILAF
Heat oil in a large saucepan on medium heat. Cook onions until translucent. Add garlic plus a pinch of salt and pepper, cook for 1 minute. Add mushrooms and cook until soft. Stir in rice and cook for 3 minutes. Add broth, turn heat to high and bring to a boil. Cover, reduce heat to simmer and cook for 15 minutes or until tender but not done.

Add almonds, replace cover and cook for 5 to 10 minutes or until done. Season to taste with salt and pepper. Garnish with parsley.

THE SALMON
Season salmon with salt and freshly ground pepper. Set a sauté pan on medium-high and heat olive oil. Place the salmon in the oil and reduce heat to medium. Once a golden crust has formed, (4-6 minutes) flip the salmon and cook for additional 2-3 minutes, until the fish flakes easily when tested with a fork.

To serve, divide rice pilaf between two plates. Place a salmon fillet on each plate, sprinkle with parsley and enjoy.

Note: *Local fish and seafood purchased in season tastes best and is your best buy for the money. The trick is to be flexible and watch for specials at your local supermarket.*

MAKES 2 SERVINGS

Ole & me running to make a set. Love the beautiful flat calm days.

ROASTED RED PEPPER AND SALMON LASAGNA

INGREDIENTS
5 large red bell peppers
12 to 14 lasagna noodles
2 (14-to 15-ounce) cans salmon, drained
½ cup fresh dill, chopped
2 tablespoons lemon juice
¼ cup plus 5 tablespoons olive oil
salt and freshly ground black pepper
4 cloves garlic
1 tablespoon sherry vinegar
4 cups watercress, rinsed
½ cup fresh chives, chopped

A delicate harmony of color distinguishes this attractive cold salad. This is a nice light meal for a hot summer day.

Place bell peppers on a baking sheet under a preheated broiler and broil, turning every 5 minutes until blackened on all sides. Transfer to a bowl and cover. Let sweat for about 10 minutes. Peel, stem and seed the peppers.

Bring a large pot of water to a boil, lightly salt it, add the pasta and cook until al dente. Drain and rinse with cold water. Place in a single layer on a parchment lined baking sheet to prevent sticking.

In a medium bowl, combine the salmon, dill, lemon juice and ¼-cup olive oil, season with salt and pepper.

Using a blender; puree the roasted red peppers, garlic, vinegar and ½ teaspoon salt. Pour in remaining 5 tablespoons olive oil, blending until thick.

Place a layer of lasagna noodles in the bottom of a 9 by 13-inch baking dish. Top with one-third of the salmon mixture, one-third of the watercress and one-quarter of the bell pepper sauce. Repeat 2 more times, using remaining salmon mixture, watercress and topping with the remaining noodles.

To serve, divide the lasagna into portions. Top with remaining red pepper sauce and the chives.

MAKES 6 SERVINGS

Fishermen's back porch.

The F/V Caron offloading to the
processing plant in Ketchikan, Alaska

GALLEY TREATS

Save room for dessert, because you won't want to miss a bite of these luscious, satisfying treats.

LEMON MERINGUE PIE

NANAIMO BARS

COOKIE CRUST FUDGE PIE

ALMOST A SIN DESSERT

BROWNIE CHEESECAKE TORTE

BANANA CREAM PIE WITH A
PEANUT BUTTER COOKIE CRUST

SWEET POTATO PECAN PIE

CHOCOLATE MOCHA BAKED ALASKA

PROCESSING PLANT
WHERE THE SALMON ARE PREPARED FOR MARKET

Once the tender or a fishing boat returns from the fishing grounds, the salmon are offloaded right away at the processing plant to maintain high quality. Depending on how the fish is used, processing methods vary. Fish for the fresh market are frozen whole or filleted and flash frozen. If smoked the fish is filleted and taken to the smoker. If canned the fish are headed, finned, gutted and hand packed into the can and put into the cookers.

From the processing plant the processed fish are packed into containers and shipped to supermarkets and fresh markets all over the world.

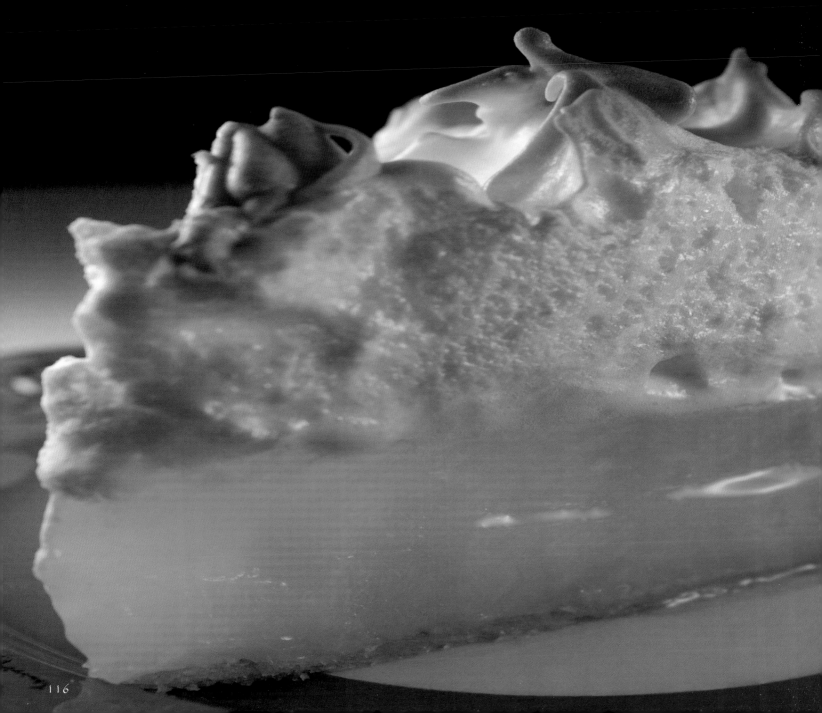

LEMON MERINGUE PIE

INGREDIENTS
1 (9-inch) pie shell, baked
1 ⅓ cups, plus ⅓ cup sugar
6 tablespoons cornstarch
½ cup freshly squeezed lemon juice
4 eggs separated
1 ½ cups boiling water
2 tablespoons butter
¼ teaspoon cream of tartar
mint leaves, optional

We all have so many fond memories of Kay's Kitchen in Ketchikan, Alaska. The famous peanut butter pie, apple pie, cherry cream cheese. For me it was the luscious lemon meringue pie. This is a fun and easy recipe to follow, because Kay made her pies sweet, simple and delicious!

Preheat the oven to 350°F.

In a heavy saucepan, combine 1 ⅓ cups sugar and cornstarch; add the lemon juice. In a small bowl, beat egg yolks; add to the lemon mixture. Gradually add the boiling water, stirring constantly. Over medium heat, cook and stir until mixture boils and thickens, about 8 minutes. Remove from heat. Add butter; stir until melted. Pour into prepared pie shell.

In a small mixer bowl, beat egg whites with cream of tartar until soft peaks form; gradually add remaining ⅓ cup sugar, beating until stiff but not dry. Spread on top of pie, sealing carefully to edge of shell. Bake 12 to 15 minutes or until golden brown. Cool. Chill before serving. Garnish with mint if desired. Refrigerate leftovers.

MAKES ONE 9-INCH PIE

Seiners hard at it.

NANAIMO BARS

INGREDIENTS
FOR THE BOTTOM LAYER
½ cup unsalted butter, softened
¼ cup sugar
⅓ cup unsweetened cocoa powder
1 large egg, beaten
1 teaspoon vanilla extract
2 cups graham cracker crumbs
1 cup sweetened coconut
½ cup walnuts, chopped

FOR THE MIDDLE LAYER
¼ cup unsalted butter, softened
2-3 tablespoons milk or cream
2 tablespoons instant vanilla pudding
   powder
½ teaspoon vanilla extract
2 cups confectioner's sugar

FOR THE TOP LAYER
4 ounces semisweet chocolate
1 tablespoon unsalted butter

Over the years, Ole and I have taken our fishing boat north through the Inside Passage. One of the fun stops along the way is a town called Nanaimo in British Columbia. Many of the local coffee and donut shops along commercial street sell these creamy chocolate treats. I'm hooked on this slice of heaven, one bite and you will be too!

Butter an 8 by 8 inch baking dish. Set aside.

THE BOTTOM LAYER
In a saucepan over low heat, melt the butter. Stir in the sugar and cocoa powder and then gradually whisk in the beaten egg. Cook, stirring constantly, until mixture thickens (1 - 2 minutes). Remove from heat and stir in the vanilla extract, graham cracker crumbs, coconut and chopped nuts. Press the mixture evenly onto the bottom of the prepared dish. Cover and refrigerate until firm, about an hour.

THE MIDDLE LAYER
In your electric mixer or with a hand mixer, beat the butter until smooth and creamy. Add the remaining ingredients and beat until the mixture is smooth. If the mixture is too thick to spread, add a little more milk. Spread the filling over the bottom layer, cover and refrigerate until firm, about 30 minutes.

THE TOP LAYER
In a heat-proof bowl over a saucepan of simmering water, melt the chocolate and butter. Spread the melted chocolate evenly over the filling and refrigerate until the chocolate has set.

To serve, to prevent the chocolate from cracking, using a sharp knife, bring the squares to room temperature before cutting. Enjoy!

MAKES ABOUT 25 SQUARES

Beach combed treasures.

COOKIE CRUST FUDGE PIE

INGREDIENTS
1 cup butter, softened
1 cup sugar
¼ cup baking cocoa
¼ cup all-purpose flour
2 eggs, beaten
1 teaspoon vanilla extract
1 teaspoon instant coffee granules
1 tablespoon coffee flavored liqueur
Pinch of salt
1 cup pecans, chopped
1 Oreo® cookie pie shell
Garnish: whipped cream, chocolate
   shavings or chocolate curls

While fishing in Alaska, this is my go to dessert recipe. It's simple, yummy and is our version of a big chocolate chip cookie.

Preheat the oven to 375°F.

In a large bowl, combine the butter, sugar, baking cocoa and flour. Mix well. Add eggs, vanilla, coffee granules, coffee liqueur, salt and pecans and mix well. Pour mixture into pie shell.

Bake for 25 to 30 minutes. Do not over bake; filling will not be firm in the center. Slice and serve warm. Top with whipped cream and chocolate shavings or chocolate curls, if desired.

MAKES 6 SERVING

ALMOST A SIN DESSERT

INGREDIENTS
FOR THE FIRST LAYER

1 **cup butter, softened**
1 **cup pecans, chopped**
2 **cups all purpose flour**

FOR THE SECOND LAYER

1 **cup confectioners sugar**
1 **(8-ounce) package cream cheese, softened**

FOR THE THIRD LAYER

2 **small packages of instant vanilla pudding**
3 **cups milk**

FOR THE FOURTH LAYER

2 **small packages instant chocolate pudding**
3 **cups milk**

1 **(8-ounce) carton of whipped topping**
shaved chocolate

I believe that dessert should be special, for when you want to treat yourself especially well. When indulgence is occasional, it should be exquisite. With that in mind, this dessert is well worth it!

Preheat the oven to 325°F.

THE FIRST LAYER
Mix together the butter, pecans and flour. Pat down into a 9 by 13 inch baking dish. Bake for 15 to 20 minutes until lightly brown. Let cool completely.

THE SECOND LAYER
Mix together the confectioners sugar and the cream cheese. Spread over completely cooled crust.

THE THIRD LAYER
Mix together the pudding and the milk, following package directions. Pour over second layer.

THE FOURTH LAYER
Mix together the pudding and the milk, following package directions. Pour over third layer.

Frost with whipped topping. Sprinkle shaved chocolate and chopped pecans over the top if desired.

Refrigerate.

SERVING SIZE:
THE LORD NEVER GIVES YOU
MORE THAN YOU CAN HANDLE.

Salmonberries.

Brownie cheesecake torte

INGREDIENTS
- 1 (18-ounce) package fudge brownie mix
- 2 teaspoons instant coffee granules
- ½ teaspoon ground cinnamon
- 1 (4-ounce) jar carrot baby food
- Cooking spray
- ½ cup plus 2 tablespoons sugar, divided
- 4 teaspoons all-purpose flour
- 1 teaspoon vanilla extract
- 1 (8-ounce) package Neufchâtel cheese, softened
- 1 (8-ounce) package cream cheese, softened
- 2 large egg whites
- 3 tablespoons milk, divided
- 2 tablespoons unsweetened cocoa
- chocolate syrup (optional)
- fresh raspberries (optional)

This cheesecake is dramatic and decadent, a real crowd-pleaser. You may be wondering, why the puréed carrots? Brownies can be a bit dry, the carrots add moisture and extra flavor.

Preheat the oven to 350°F.

Combine the fudge brownie mix, coffee granules, cinnamon and puréed carrots in a bowl. Firmly press mixture into bottom and 1 inch up sides of a 9-inch spring form pan coated with cooking spray. Set aside.

Combine ½ cup sugar, flour, vanilla and cheeses; beat at low speed with an electric mixer until well-blended. Gently add the egg whites and 2 tablespoons milk. (Take special care not to whip air into the batter, as it will cause the cheesecake to puff in the oven and then fall when removed).

Combine ½ cup batter, 1 tablespoon milk, 2 tablespoons sugar and cocoa in a small bowl; stir well. Spoon remaining batter alternately with cocoa mixture into prepared crust. Swirl together using the tip of a knife.

Bake for 10 minutes. Reduce oven temperature to 250°F; bake 45 minutes or until almost set. Cool completely on a wire rack. Garnish with chocolate syrup and fresh raspberries, if desired.

__Note:__ Cheesecakes are among the most frequently overcooked foods, they are the most deceptive when trying to figure out when they are done baking. When it's done, it never "looks" done. To test if a cheesecake is done baking, gently shake the pan. The top of the cake should move as one solid piece, but its center should still be wobbly (not soupy) in about a 3-inch circle in the center.

MAKES 6-8 SERVINGS

Near the south end of the Stikine Icefield, Leconte Glacier is the southernmost active tidewater glacier in the northern hemisphere.

BANANA CREAM PIE WITH A
PEANUT BUTTER COOKIE CRUST

INGREDIENTS
½ lb. peanut butter sandwich cookies,
    coarsely crumbled
¼ teaspoon salt
2 tablespoons butter, melted
2 eggs
1 tablespoon rum, optional
1 cup sugar
2 tablespoons cornstarch
1 tablespoon all-purpose flour
Pinch of salt
2 cups whole milk
2 tablespoons unsalted butter
½ teaspoon vanilla extract
3 ripe, firm bananas, quartered
whipped cream

*Note: Chill pie overnight, then remove
sides of pan. To slice, dip a knife in hot
water before making cuts, rinsing the
blade clean each time.*

*Banana cream pie is one of the yummiest things on earth and this is
a great one!*

Coat a 9" tart pan with removable bottom with nonstick spray.

Process cookies and salt for the crust in a food processor until fine. Drizzle in
butter while machine is running. Transfer crumbs to prepared pan and lightly
press to cover bottom and sides; set aside.

Whisk eggs and rum for the filling in a small bowl.

Combine sugar, cornstarch, flour and salt in a saucepan, whisking to break up
lumps. Gradually add milk and whisk until smooth. Cook over medium heat
until thickened, about 8 minutes. Temper some of the hot custard into the egg
mixture, then add the egg mixture to remaining custard. Continue to cook until
thick and bubbly, about 3 minutes. Remove from heat.

Add unsalted butter and vanilla. Assemble pie by arranging bananas in the
crust, cut side down. Pour filling over bananas and smooth to cover. Wrap
loosely in plastic and chill overnight.

Serve pie plain or with whipped cream.

MAKES ONE 9" PIE

Wild daisies.

Huckleberry blossom.

SWEET POTATO PECAN PIE

INGREDIENTS
1 (9-inch) frozen unbaked pie shell
1 cup sweet potatoes, mashed
2 tablespoons brown sugar
2 large egg yolks
1 tablespoon butter, softened
1 tablespoon heavy cream
1½ teaspoons vanilla extract, divided
¼ plus ⅛ teaspoon ground cinnamon
¼ teaspoon salt
⅛ teaspoon ground nutmeg
1 cup pecans, chopped
½ cup pecans, halved
¼ cup sugar
2 large eggs
¾ cup light corn syrup
1½ tablespoons butter, melted
whipped topping
ice cream

Ditch the hassle of choosing between sweet potato and pecan pie. With this simple recipe, you can have both in each bite.

Preheat the oven to 400°F.

Bake pie shell 6 to 8 minutes or until lightly browned; remove from oven. Lower temperature to 350°F.

Combine the sweet potato, brown sugar, egg yolks, butter and cream in a large bowl. Add ½ teaspoon vanilla, cinnamon, salt and nutmeg. Stir until smooth and completely combined.

Spread in an even layer in bottom of pie shell. Add chopped pecans evenly over sweet potato mixture. Lay pecan halves flat in a circle around perimeter.

Combine ¼ cup sugar and 2 large eggs in a separate bowl; whisk until blended. Whisk in corn syrup, melted butter and remaining 1 teaspoon vanilla. Pour mixture slowly over pecan layer. Do not over fill. (You may have excess).

Bake at 350°F. for 50 to 55 minutes or until pie is puffed, set and golden brown. Serve with whipped topping or ice cream.

Note: *One large, baked sweet potato-should yield a little more than 1 cup purée. Bake until soft (about 45 minutes), remove skin and mash with a potato masher until smooth. Should you prefer to use canned sweet potatoes, be sure to drain them well.*

MAKES 6 SERVINGS

The F/V Bronze Maiden.

CHOCOLATE MOCHA BAKED ALASKA

INGREDIENTS

8 ounces semi sweet chocolate, broken into pieces
1 cup heavy cream
1 purchased pound cake, (about 14 oz.)
6 tablespoons coffee liqueur, (or 4 Tbl. strong coffee)
1 quart coffee ice cream
muffin pan
muffin liners
chill baking sheet in the freezer

FOR THE MERINGUE

6 egg whites, room temperature
½ teaspoon cream of tartar
⅛ teaspoon salt
1 cup sugar
1 teaspoon vanilla extract

Note: The name Baked Alaska originated at Delmonico's Restaurant in New York City in 1876 and was created in honor of the newly acquired territory of Alaska. February 1st is Baked Alaska Day.

Preparation Tips: Assembling Alaskas ahead (as little as a day or up to a week) insures the ice cream is frozen solid. This is important so it can withstand the heat. A Jumbo muffin pan, custard cups and ramekins (small ceramic bowls) also work great for forming individual portions. Line it with pieces of plastic wrap.

MAKES 12 SERVINGS

The secret to this wonder is in its construction. Cake and ice cream are frozen together, then covered with meringue. The meringue acts as insulation, protecting the ice cream from the heat. The cake acts as a buffer from below and shields the ice cream. Baked Alaska is truly a unique treat. Your family will love it!

Heat chocolate and cream in a heavy small saucepan. Whisk over low heat until mixture is smooth. Chill 30 minutes until spreadable. Reserve extra ganache (icing) for plating.

Line muffin pan with muffin liners or pieces of plastic wrap, leaving overhang.

Prepare the pound cake, slicing it horizontally into four layers. Cut rounds from each layer; matching the size of your muffin pan openings.

Press ¼ cup ice cream into the lined muffin tin. Leave ½ inch space on top for the cake round. Pour 1 teaspoon liqueur (or 2 teaspoons coffee) onto each round, then spread with 2 tablespoons ganache. Place a cake round on top of the ice cream, ganache side down. Cover the pan tightly with plastic wrap and freeze until firm, at least 4 hours.

THE MERINGUE

Beat egg whites in a mixing bowl until foamy. Add cream of tartar and salt. Beat at medium-high speed until the whites hold a soft peak. Beat in the vanilla. Gradually add the sugar 1 tablespoon at a time. Continue beating until stiff glossy peaks form. (The meringue needs to be sturdy to form "insulation" and keep the ice cream from melting).

TO ASSEMBLE

Line the frozen sheet pan with parchment paper. Place the Alaskas on the parchment, cake side down. Quickly, remove the cupcake liners from the frozen ice cream. Generously cover each Alaska with 1 cup of meringue, spreading to form cloud-like swirls. Be sure the meringue reaches the parchment, creating a barrier so the ice cream doesn't leak. Place in freezer until ready to bake.

To bake: preheat the oven to 500°F. Take Alaskas directly from the freezer to the oven with rack in the top third. Bake 2 to 3 minutes just until the tips are lightly browned. If a little ice cream leaks, don't worry about it. Carefully transfer Alaskas to chilled plates. Spoon warm ganache around dessert and serve.

129

How do we do it 24/7?
A Relationship Recipe

Is it possible to work together and still maintain the spark that brought us together as a couple?

Yes! 24/7 works-when you know what to do. I fell in love with a fisherman. To each other we were just two people madly in love. We went to dinner, the movies and walked up and down boat harbors. We held hands, we kissed and we could not stand to be apart for more than a few minutes. We were head over heels in love. "Wouldn't it be great if we bought a fishing business and fished together?" "Then we wouldn't have to be apart!" It was a great idea. We would merge our talents and skills to create a fishing business. We were going to be successful, in love and together 24 hours a day, 7 days a week for the rest of our lives. We could not stop talking about our new adventure. We sat for hours dreaming of our new boat and looking through boating magazines. Soon our walks down to the boat harbors became a search for the perfect fishing boat. Our cozy romantic dinners included number crunching and legal pads. Our pillow talk was about the areas we would fish. On a trip to Wrangell, Alaska we found our first boat, the fishing vessel Kathryn Ann.

We were now much more than romantic partners, we were business partners. Our fishing days in the beginning were long and hard, but we were together for every waking moment of them. As our fishing business expanded to include pot shrimping and long lining halibut, we were focused on catching fish and not on each other. There simply was not enough time or energy in the day. It was time to divide and conquer. I would handle all of the shopping, cooking, cleaning and the business end of fishing. Ole would catch as many fish as he could and keep the boat running smoothly; after all, the goal was success. A marriage is like a house - the more time you spend in it, the more lived-in and dustier it becomes. Of course you don't always see the normal wear and tear of your house - sometimes you're just too busy.

Like our boat, each fall when the fishing season ends, we give our marriage a detailed inspection. We find areas that worked well and what didn't work so well. We figure out what needs a little tweaking. We set this as a high priority in our marriage and take the time to work on this together. The first place we start is communication. We take a close look at how we communicate with each other. When Ole is out on deck and our net is hung up under the boat full of fish, it's not a good time to ask if we can go to town. Ole knows if I'm in the galley trying to put a meal together and it's rough out. He does his best to keep the boat out of the trough. (The trough is the bottom of the wave. If the boat lays sideways in the trough, the boat rolls twice as much).

New Eddystone Rock, Behm Canal.

We learned to become efficient in our communications, to leave out details and rambling and present the high points to each other. We request, rather than command, we value each other's opinion and include, rather than dictate and we never ever assume we know what the other person is feeling or will say. Our top two goals are "time-saving" and conflict avoidance. We talk nicely "with" each other so that our day is productive and respectful of our work relationship. The most complicated part to learning this was realizing that we both are going to have to change if we were going to find new ways to communicate. Ole needed to learn to open up and talk a little more throughout the day and I needed to learn to be content and silent.

In a standard marriage you have 8 to 10 hours apart from each other when you are at work. During that time, you have an opportunity to laugh off the little things that irked you that morning. You can blow off steam around the water cooler with a friend. By the time you get home, you have long forgotten that little thing and can't wait to happily see each other. In a 24/7 fishing marriage, you do not get that space. There is no time or space to distance yourself from the problem and you certainly cannot jump in the water, swim to shore and talk about it with a friend. To compound it all, we are both equally sleep-deprived from the 20 hour work days and tend to be grumpy, tired and step on each other's toes all day long. "So, what do you want for dinner?" "I don't care, whatever you want is fine," Ole replies. That is the final straw. All the things we have held in for the entire week come screaming to the surface. Both of us are tired and it becomes who can top the other with tales of errors in behavior, judgment and memory loss. There seems to be no end to the list of things we have been stockpiling.

We made an agreement long ago to discuss the little things so they will not pile up. When we first agreed to work together, we had a picture of what it would be like and how we would use our free time together. We took the time to design our lives together instead of reacting to each event as it occurs. We live our lives creating new and exciting opportunities and never stop saying, "Is this how we want to live?" We have devoted our marriage to doing something we both enjoy equally and that is fishing. We realize that we have a different kind of stress than other married couples and the fastest cure for that stress is laughter. We laugh together daily. We laugh when we are at home and we laugh when we are fishing. In laughing we find the love that keeps our marriage and our relationship alive.

As I write the words on this page, I realize how lucky Ole and I are and how precious the life we have together truly is 24/7.

This is our girl, a female Tabby named **Sobe**.

Sobe is a sweet gal with a big heart and likes her cuddles. She is so innocent, curious and adventurous in every way imaginable. Although she is mostly an inside kitty, she does enjoy her summers on the boat. She isn't much of a hunter. She would rather relax and enjoy the wonderful smells the ocean breeze brings, than bother to chase anything other than her furry little ball.

A couple of years ago Ole and I finished up an intense week of fishing. With a 24 hour rest period in between openers, we were catching up on our much needed sleep.

Both sound asleep in a quiet harbor, we were awakened by a commotion - as we focused our eyes we both saw Sobe up on the dashboard with a tiny wild bird in her mouth. She had this look on her face like "Oh my gosh, did I really just catch my first bird?" and "What do I do with it?" We looked at the tiny bird that was being held ever so gently...(to this day we both swear the bird made eye contact with us pleading for mercy). With a strong but gentle voice we commanded Sobe to let go of the bird! Sobe did let go and with feathers flying, the tiny bird exited out the window! We rewarded Sobe with crab, satisfied she found her way back to the dashboard for a cat nap.

Sobe is the perfect boat cat. She has been with us up and down the coast in the fiercest of storms, even Queen Charlotte Sound in the storm of the century! If she could talk, I wonder the stories she would tell.

INDEX OF RECIPES